MW00962184

VECTOR CALCULUS FOR ENGINEERS

Vector Calculus for Engineers

Jeffrey Robert Chasnov

Professor of Mathematics
The Hong Kong University
of Science and Technology

The Hong Kong University of Science and Technology
Department of Mathematics
Clear Water Bay, Kowloon
Hong Kong

Copyright © 2022 by Jeffrey Robert Chasnov. All rights reserved.

Contents

Preface

This is the paperback edition of the lecture notes for my online Coursera course, ***Vector Calculus for Engineers***. I have divided these notes into chapters called Lectures, with each Lecture corresponding to a video on Coursera.

There are problems at the end of each lecture chapter and I have tried to choose problems that exemplify the main idea of the lecture. Students taking a formal university course in Vector Calculus (sometimes called Multivariable Calculus or Calc 3) will usually be assigned many more additional problems, but here I follow the philosophy that less is more. I give enough problems for students to solidify their understanding of the material, but not so many problems that students feel overwhelmed and drop out. I do encourage students to attempt the given problems, but if they get stuck, full solutions can be found in the Appendix.

There are also additional problems at the end of coherent sections that are given as practice quizzes on the Coursera platform. Again, students should attempt these quizzes on the platform, but if a student has trouble obtaining a correct answer, full solutions are also found in the Appendix.

Students who take this course are expected to already know single-variable differential and integral calculus to the level of a first-year college calculus course. Students should also be familiar with matrices, and be able to compute a three-by-three determinant.

JEFFREY R. CHASNOV
Hong Kong
Nov 2022

Week I

Vectors

In this week's lectures, we learn about vectors. Vectors are line segments with both length and direction, and are fundamental to engineering mathematics. We will define vectors, how to add and subtract them, and how to multiply them using the scalar and vector products (dot and cross products). We use vectors to learn some analytical geometry of lines and planes, and introduce the Kronecker delta and the Levi-Civita symbol to prove vector identities. The important concepts of scalar and vector fields are discussed.

Lecture 1 Vectors

We define a vector in three-dimensional Euclidean space as having a length (or magnitude) and a direction. A vector is depicted as an arrow starting at one point in space and ending at another point. All vectors that have the same length and point in the same direction are considered equal, no matter where they are located in space. (Variables that are vectors will be denoted in print by boldface, and in hand by an arrow drawn over the symbol.) In contrast, scalars have magnitude but no direction. Zero can either be a scalar or a vector and has zero magnitude. The negative of a vector reverses its direction. Examples of vectors are velocity and acceleration; examples of scalars are mass and charge.

Vectors can be added to each other and multiplied by scalars. A simple example is a mass m acted on by two forces $\boldsymbol{F}_1$ and $\boldsymbol{F}_2$. Newton's equation then takes the form $m\boldsymbol{a} = \boldsymbol{F}_1 + \boldsymbol{F}_2$, where $\boldsymbol{a}$ is the acceleration vector of the mass. Vector addition is commutative and associative:

$$\boldsymbol{A} + \boldsymbol{B} = \boldsymbol{B} + \boldsymbol{A}, \qquad (\boldsymbol{A} + \boldsymbol{B}) + \boldsymbol{C} = \boldsymbol{A} + (\boldsymbol{B} + \boldsymbol{C});$$

and scalar multiplication is distributive:

$$k(\boldsymbol{A} + \boldsymbol{B}) = k\boldsymbol{A} + k\boldsymbol{B}.$$

Multiplication of a vector by a positive scalar changes the length of the vector but not its direction. Vector addition can be represented graphically by placing the tail of one of the vectors on the head of the other. Vector subtraction adds the first vector to the negative of the second. Notice that when the tail of $\boldsymbol{A}$ and $\boldsymbol{B}$ are placed at the same point, the vector $\boldsymbol{B} - \boldsymbol{A}$ points from the head of $\boldsymbol{A}$ to the head of $\boldsymbol{B}$, or equivalently, the tail of $-\boldsymbol{A}$ to the head of $\boldsymbol{B}$.

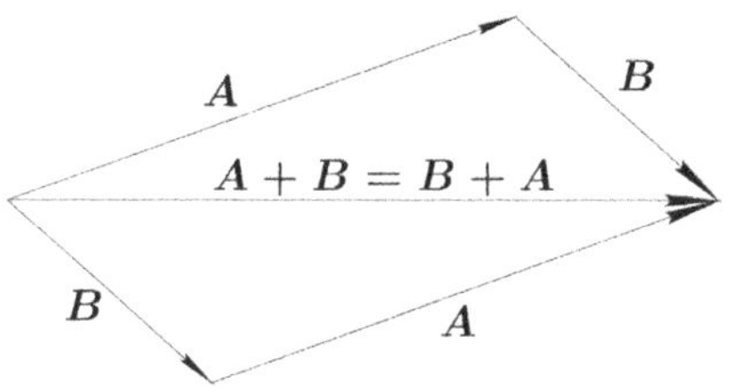

Vector addition

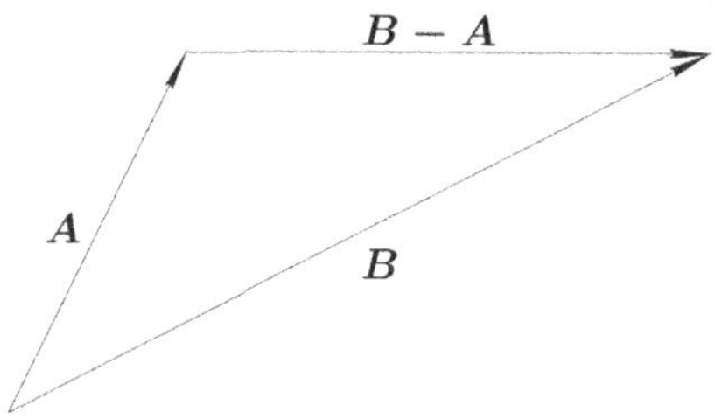

Vector subtraction

Problems for Lecture 1

1. Show graphically that vector addition is associative, that is, $(\boldsymbol{A}+\boldsymbol{B})+\boldsymbol{C}=\boldsymbol{A}+(\boldsymbol{B}+\boldsymbol{C})$.

2. Using vectors, prove that the line segment joining the midpoints of two sides of a triangle is parallel to the third side and half its length.

Lecture 2 | Cartesian coordinates

To solve a physical problem, we usually impose a coordinate system. The familiar three-dimensional x-y-z coordinate system is called the Cartesian coordinate system. Three mutually perpendicular lines called axes intersect at a point called the origin, denoted as $(0,0,0)$. All other points in three-dimensional space are identified by their coordinates as (x,y,z) in the standard way. The positive directions of the axes are usually chosen to form a right-handed coordinate system. When one points the right hand in the direction of the positive x-axis, and curls the fingers in the direction of the positive y-axis, the thumb should point in the direction of the positive z-axis.

A vector has a length and a direction. If we impose a Cartesian coordinate system and place the tail of a vector at the origin, then the head points to a specific point. For example, if the vector $\boldsymbol{A}$ has head pointing to (A_1, A_2, A_3), we say that the x-component of $\boldsymbol{A}$ is A_1, the y-component is A_2, and the z-component is A_3. The length of the vector $\boldsymbol{A}$, denoted by $|\boldsymbol{A}|$, is a scalar and is independent of the orientation of the coordinate system. Application of the Pythagorean theorem in three dimensions results in

$$|\boldsymbol{A}| = \sqrt{A_1^2 + A_2^2 + A_3^2}.$$

We can define standard unit vectors $\boldsymbol{i}$, $\boldsymbol{j}$ and $\boldsymbol{k}$, to be vectors of length one that point along the positive directions of the x-, y-, and z-coordinate axes, respectively. Using these unit vectors, we will write a vector as

$$\boldsymbol{A} = A_1\boldsymbol{i} + A_2\boldsymbol{j} + A_3\boldsymbol{k}.$$

With also $\boldsymbol{B} = B_1\boldsymbol{i} + B_2\boldsymbol{j} + B_3\boldsymbol{k}$, vector addition and scalar multiplication can be expressed component-wise and is given by

$$\boldsymbol{A}+\boldsymbol{B} = (A_1+B_1)\boldsymbol{i} + (A_2+B_2)\boldsymbol{j} + (A_3+B_3)\boldsymbol{k}, \quad c\boldsymbol{A} = cA_1\boldsymbol{i} + cA_2\boldsymbol{j} + cA_3\boldsymbol{k}.$$

The position vector, $\boldsymbol{r}$, is defined as the vector that points from the origin to the point (x,y,z), and is used to locate a specific point in space. It can be written in terms of the standard unit vectors as

$$\boldsymbol{r} = x\boldsymbol{i} + y\boldsymbol{j} + z\boldsymbol{k}.$$

A displacement vector is the difference between two position vectors. For position vectors $\boldsymbol{r}_1$ and $\boldsymbol{r}_2$, the displacement vector that points from the head of $\boldsymbol{r}_1$ to the head of $\boldsymbol{r}_2$ is given by

$$\boldsymbol{r}_2 - \boldsymbol{r}_1 = (x_2 - x_1)\boldsymbol{i} + (y_2 - y_1)\boldsymbol{j} + (z_2 - z_1)\boldsymbol{k}.$$

Problems for Lecture 2

1. Given a Cartesian coordinate system with standard unit vectors $\boldsymbol{i}$, $\boldsymbol{j}$, and $\boldsymbol{k}$, let a mass m_1 be at position $\boldsymbol{r}_1 = x_1\boldsymbol{i} + y_1\boldsymbol{j} + z_1\boldsymbol{k}$ and a mass m_2 be at position $\boldsymbol{r}_2 = x_2\boldsymbol{i} + y_2\boldsymbol{j} + z_2\boldsymbol{k}$. In terms of the standard unit vectors, determine the unit vector that points from m_1 to m_2.

2. Newton's law of universal gravitation states that two point masses attract each other along the line connecting them, with a force proportional to the product of their masses and inversely proportional to the square of the distance between them. The magnitude of the force acting on each mass is therefore

$$F = G\frac{m_1 m_2}{r^2},$$

where m_1 and m_2 are the two masses, r is the distance between them, and G is the gravitational constant. Let the masses m_1 and m_2 be located at the position vectors $\boldsymbol{r}_1$ and $\boldsymbol{r}_2$. Write down the vector form for the force acting on m_1 due to its gravitational attraction to m_2.

Lecture 3 Dot product

We define the dot product (or scalar product) between two vectors $\boldsymbol{A} = A_1\boldsymbol{i} + A_2\boldsymbol{j} + A_3\boldsymbol{k}$ and $\boldsymbol{B} = B_1\boldsymbol{i} + B_2\boldsymbol{j} + B_3\boldsymbol{k}$ as

$$\boldsymbol{A} \cdot \boldsymbol{B} = A_1B_1 + A_2B_2 + A_3B_3.$$

One can prove that the dot product is commutative, distributive over addition, and associative with respect to scalar multiplication; that is,

$$\boldsymbol{A} \cdot \boldsymbol{B} = \boldsymbol{B} \cdot \boldsymbol{A}, \qquad \boldsymbol{A} \cdot (\boldsymbol{B} + \boldsymbol{C}) = \boldsymbol{A} \cdot \boldsymbol{B} + \boldsymbol{A} \cdot \boldsymbol{C},$$

$$\boldsymbol{A} \cdot (c\boldsymbol{B}) = (c\boldsymbol{A}) \cdot \boldsymbol{B} = c(\boldsymbol{A} \cdot \boldsymbol{B}).$$

A geometric interpretation of the dot product is also possible. Given any two vectors $\boldsymbol{A}$ and $\boldsymbol{B}$, place the vectors tail-to-tail, and impose a coordinate system with origin at the tails such that $\boldsymbol{A}$ is parallel to the x-axis and $\boldsymbol{B}$ lies in the x-y plane, as shown in the figure. The angle between the two vectors is denoted as θ.

Then in this coordinate system, $\boldsymbol{A} = |\boldsymbol{A}|\boldsymbol{i}$, $\boldsymbol{B} = |\boldsymbol{B}|\cos\theta\boldsymbol{i} + |\boldsymbol{B}|\sin\theta\boldsymbol{j}$, and

$$\boldsymbol{A} \cdot \boldsymbol{B} = |\boldsymbol{A}||\boldsymbol{B}|\cos\theta,$$

a result independent of the choice of coordinate system. If $\boldsymbol{A}$ and $\boldsymbol{B}$ are parallel, then $\theta = 0$ and $\boldsymbol{A} \cdot \boldsymbol{B} = |\boldsymbol{A}||\boldsymbol{B}|$ and in particular, $\boldsymbol{A} \cdot \boldsymbol{A} = |\boldsymbol{A}|^2$. If $\boldsymbol{A}$ and $\boldsymbol{B}$ are perpendicular, then $\theta = \pi/2$ and $\boldsymbol{A} \cdot \boldsymbol{B} = 0$.

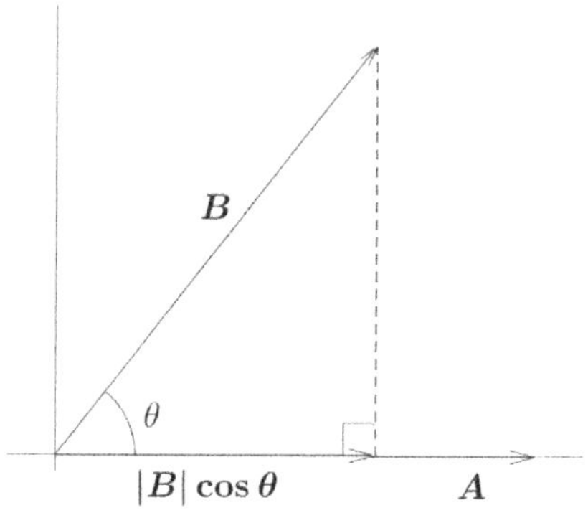

Problems for Lecture 3

1. Using the definition of the dot product $\boldsymbol{A} \cdot \boldsymbol{B} = A_1B_1 + A_2B_2 + A_3B_3$, prove that

a) $\boldsymbol{A} \cdot \boldsymbol{B} = \boldsymbol{B} \cdot \boldsymbol{A}$;

b) $\boldsymbol{A} \cdot (\boldsymbol{B} + \boldsymbol{C}) = \boldsymbol{A} \cdot \boldsymbol{B} + \boldsymbol{A} \cdot \boldsymbol{C}$;

c) $\boldsymbol{A} \cdot (k\boldsymbol{B}) = (k\boldsymbol{A}) \cdot \boldsymbol{B} = k(\boldsymbol{A} \cdot \boldsymbol{B})$.

2. Determine all the combinations of dot products between the standard unit vectors $\boldsymbol{i}$, $\boldsymbol{j}$, and $\boldsymbol{k}$.

3. Let $\boldsymbol{C} = \boldsymbol{A} - \boldsymbol{B}$. Calculate the dot product of $\boldsymbol{C}$ with itself and thus derive the law of cosines.

Lecture 4 | Cross product

We define the cross product (or vector product) between the following two vectors $\boldsymbol{A} = A_1\boldsymbol{i} + A_2\boldsymbol{j} + A_3\boldsymbol{k}$ and $\boldsymbol{B} = B_1\boldsymbol{i} + B_2\boldsymbol{j} + B_3\boldsymbol{k}$ as

$$\begin{aligned} \boldsymbol{A} \times \boldsymbol{B} &= \begin{vmatrix} \boldsymbol{i} & \boldsymbol{j} & \boldsymbol{k} \\ A_1 & A_2 & A_3 \\ B_1 & B_2 & B_3 \end{vmatrix} \\ &= (A_2B_3 - A_3B_2)\boldsymbol{i} + (A_3B_1 - A_1B_3)\boldsymbol{j} + (A_1B_2 - A_2B_1)\boldsymbol{k}. \end{aligned}$$

Defining the cross product by a determinant serves to remember the formula. One can prove that the cross product is anticommutative, distributive over addition, and associative with respect to scalar multiplication; that is

$$\boldsymbol{A} \times \boldsymbol{B} = -\boldsymbol{B} \times \boldsymbol{A}, \qquad \boldsymbol{A} \times (\boldsymbol{B} + \boldsymbol{C}) = \boldsymbol{A} \times \boldsymbol{B} + \boldsymbol{A} \times \boldsymbol{C},$$

$$\boldsymbol{A} \times (c\boldsymbol{B}) = (c\boldsymbol{A}) \times \boldsymbol{B} = c(\boldsymbol{A} \times \boldsymbol{B}).$$

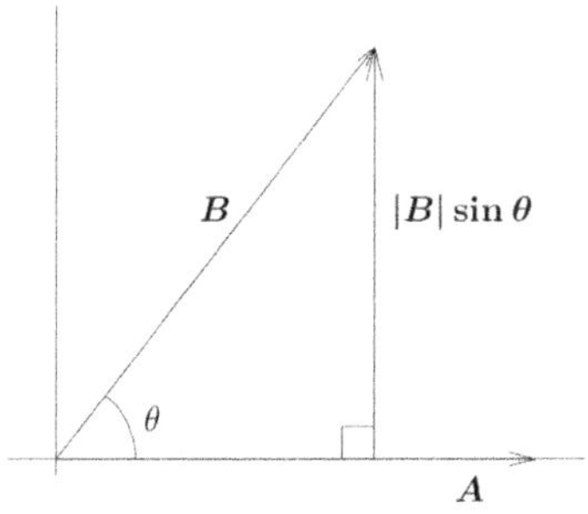

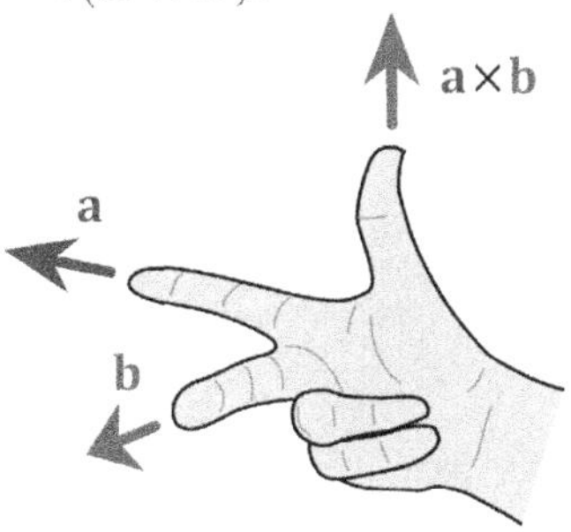

Right-hand rule (from Wikipedia)

A geometric interpretation of the cross product is also possible. Given two vectors $\boldsymbol{A}$ and $\boldsymbol{B}$ with angle θ between them, impose a coordinate system so that $\boldsymbol{A}$ is parallel to the x-axis and $\boldsymbol{B}$ lies in the x-y plane. Then $\boldsymbol{A} = |\boldsymbol{A}|\boldsymbol{i}$, $\boldsymbol{B} = |\boldsymbol{B}|\cos\theta\boldsymbol{i} + |\boldsymbol{B}|\sin\theta\boldsymbol{j}$, and $\boldsymbol{A} \times \boldsymbol{B} = |\boldsymbol{A}||\boldsymbol{B}|\sin\theta\boldsymbol{k}$. The coordinate-independent relationship is

$$|\boldsymbol{A} \times \boldsymbol{B}| = |\boldsymbol{A}||\boldsymbol{B}|\sin\theta,$$

Furthermore, the vector $\boldsymbol{A} \times \boldsymbol{B}$ points in the direction perpendicular to the plane formed by $\boldsymbol{A}$ and $\boldsymbol{B}$, and its sign is determined by the right-hand rule. Also, observe that $|\boldsymbol{A} \times \boldsymbol{B}|$ is the area of the parallelogram whose adjacent sides are the vectors $\boldsymbol{A}$ and $\boldsymbol{B}$.

Problems for Lecture 4

1. Using properties of the determinant, prove that

a) $\boldsymbol{A} \times \boldsymbol{B} = -\boldsymbol{B} \times \boldsymbol{A}$;

b) $\boldsymbol{A} \times (\boldsymbol{B} + \boldsymbol{C}) = \boldsymbol{A} \times \boldsymbol{B} + \boldsymbol{A} \times \boldsymbol{C}$;

c) $\boldsymbol{A} \times (k\boldsymbol{B}) = (k\boldsymbol{A}) \times \boldsymbol{B} = k(\boldsymbol{A} \times \boldsymbol{B})$.

2. Determine all the combinations of cross products between the standard unit vectors $\boldsymbol{i}$, $\boldsymbol{j}$, and $\boldsymbol{k}$.

3. Show that the cross product is not in general associative. That is, find an example using unit vectors such that

$$\boldsymbol{A} \times (\boldsymbol{B} \times \boldsymbol{C}) \neq (\boldsymbol{A} \times \boldsymbol{B}) \times \boldsymbol{C}.$$

Practice Quiz | Vectors

1. Let $\boldsymbol{A}$, $\boldsymbol{B}$ and $\boldsymbol{C}$ be any vectors. Which of the following statements is false?

a) $\boldsymbol{A} \cdot \boldsymbol{B} = \boldsymbol{B} \cdot \boldsymbol{A}$

b) $\boldsymbol{A} + (\boldsymbol{B} + \boldsymbol{C}) = (\boldsymbol{A} + \boldsymbol{B}) + \boldsymbol{C}$

c) $\boldsymbol{A} \times (\boldsymbol{B} \times \boldsymbol{C}) = (\boldsymbol{A} \times \boldsymbol{B}) \times \boldsymbol{C}$

d) $\boldsymbol{A} \cdot (\boldsymbol{B} + \boldsymbol{C}) = \boldsymbol{A} \cdot \boldsymbol{B} + \boldsymbol{A} \cdot \boldsymbol{C}$

2. Let $\boldsymbol{A} = a_1\boldsymbol{i} + a_2\boldsymbol{j} + a_3\boldsymbol{k}$ and $\boldsymbol{B} = b_1\boldsymbol{i} + b_2\boldsymbol{j} + b_3\boldsymbol{k}$. Then $(\boldsymbol{A} \times \boldsymbol{B}) \cdot \boldsymbol{j}$ is equal to

a) $a_2b_3 - a_3b_2$

b) $a_3b_1 - a_1b_3$

c) $a_1b_2 - a_2b_1$

d) $a_1b_3 - a_3b_1$

3. Which vector is not equal to zero?

a) $\boldsymbol{i} \times (\boldsymbol{j} \times \boldsymbol{k})$

b) $(\boldsymbol{i} \times \boldsymbol{j}) \times \boldsymbol{k}$

c) $(\boldsymbol{i} \times \boldsymbol{i}) \times \boldsymbol{j}$

d) $\boldsymbol{i} \times (\boldsymbol{i} \times \boldsymbol{j})$

Lecture 5 | Analytic geometry of lines

In two dimensions, the equation for a line in slope-intercept form is $y = mx + b$, and in point-slope form is $y - y_1 = m(x - x_1)$. In three dimensions, a line is most commonly expressed as a parametric equation.

Suppose that a line passes through a point with position vector $\boldsymbol{r}_0$ and in a direction parallel to the vector $\boldsymbol{u}$. Then, from the definition of vector addition, we can specify the position vector $\boldsymbol{r}$ for any point on the line by

$$\boldsymbol{r} = \boldsymbol{r}_0 + \boldsymbol{u}t,$$

where t is a parameter that can take on any real value.

This parametric equation for a line has clear physical meaning. If $\boldsymbol{r}$ is the position vector of a particle, then $\boldsymbol{u}$ is the velocity vector, and t is the time. In particular, differentiating $\boldsymbol{r} = \boldsymbol{r}(t)$ with respect to time results in $d\boldsymbol{r}/dt = \boldsymbol{u}$.

A nonparametric equation for the line can be obtained by eliminating t from the equations for the components. The component equations are

$$x = x_0 + u_1 t, \qquad y = y_0 + u_2 t, \qquad z = z_0 + u_3 t;$$

and eliminating t results in

$$\frac{x - x_0}{u_1} = \frac{y - y_0}{u_2} = \frac{z - z_0}{u_3}.$$

Example: Find the parametric equation for a line that passes through the points $(1, 2, 3)$ *and* $(3, 2, 1)$. *Determine the intersection point of the line with the* $z = 0$ *plane.*

To find a vector parallel to the direction of the line, we first compute the displacement vector between the two given points:

$$\boldsymbol{u} = (3 - 1)\boldsymbol{i} + (2 - 2)\boldsymbol{j} + (1 - 3)\boldsymbol{k} = 2\boldsymbol{i} - 2\boldsymbol{k}.$$

Choosing a point on the line with position vector $\boldsymbol{r}_0 = \boldsymbol{i} + 2\boldsymbol{j} + 3\boldsymbol{k}$, the parametric equation for the line is given by

$$\boldsymbol{r} = \boldsymbol{r}_0 + \boldsymbol{u}t = \boldsymbol{i} + 2\boldsymbol{j} + 3\boldsymbol{k} + t(2\boldsymbol{i} - 2\boldsymbol{k}) = (1 + 2t)\boldsymbol{i} + 2\boldsymbol{j} + (3 - 2t)\boldsymbol{k}.$$

The line crosses the $z = 0$ plane when $3 - 2t = 0$, or $t = 3/2$. At this value of t, we find $(x, y) = (4, 2)$.

Problems for Lecture 5

1. Find the parametric equation for a line that passes through the points $(1,1,1)$ and $(2,3,2)$. Determine the intersection point of the line with the $x = 0$ plane, $y = 0$ plane and $z = 0$ plane.

Lecture 6 Analytic geometry of planes

A plane in three-dimensional space is determined by three non-collinear points. Two linearly independent displacement vectors with direction parallel to the plane can be formed from these three points, and the cross-product of these two displacement vectors will be a vector that is orthogonal to the plane. We can use the dot product to express this orthogonality.

So let three points that define a plane be located by the position vectors $\boldsymbol{r}_1$, $\boldsymbol{r}_2$, and $\boldsymbol{r}_3$, and construct any two displacement vectors, such as $\boldsymbol{s}_1 = \boldsymbol{r}_2 - \boldsymbol{r}_1$ and $\boldsymbol{s}_2 = \boldsymbol{r}_3 - \boldsymbol{r}_2$. A vector normal to the plane is given by $\boldsymbol{N} = \boldsymbol{s}_1 \times \boldsymbol{s}_2$, and for any point in the plane with position vector $\boldsymbol{r}$, and for any one of the given position vectors $\boldsymbol{r}_i$, we have $\boldsymbol{N} \cdot (\boldsymbol{r} - \boldsymbol{r}_i) = 0$. With $\boldsymbol{r} = x\boldsymbol{i} + y\boldsymbol{j} + z\boldsymbol{k}$, $\boldsymbol{N} = a\boldsymbol{i} + b\boldsymbol{j} + c\boldsymbol{k}$ and $d = \boldsymbol{N} \cdot \boldsymbol{r}_i$, the equation for the plane can be written as $\boldsymbol{N} \cdot \boldsymbol{r} = \boldsymbol{N} \cdot \boldsymbol{r}_i$, or

$$ax + by + cz = d.$$

Notice that the coefficients of x, y and z are the components of the normal vector to the plane.

Example: Find an equation for the plane defined by the three points $(2, 1, 1)$, $(1, 2, 1)$, *and* $(1, 1, 2)$. *Determine the equation for the line in the x-y plane formed by the intersection of this plane with the* $z = 0$ *plane.*

To find two vectors parallel to the plane, we compute two displacement vectors from the three points:

$$\begin{aligned} \boldsymbol{s}_1 &= (1-2)\boldsymbol{i} + (2-1)\boldsymbol{j} + (1-1)\boldsymbol{k} = -\boldsymbol{i} + \boldsymbol{j}, \\ \boldsymbol{s}_2 &= (1-1)\boldsymbol{i} + (1-2)\boldsymbol{j} + (2-1)\boldsymbol{k} = -\boldsymbol{j} + \boldsymbol{k}. \end{aligned}$$

A normal vector to the plane is then found from

$$\boldsymbol{N} = \boldsymbol{s}_1 \times \boldsymbol{s}_2 = \begin{vmatrix} \boldsymbol{i} & \boldsymbol{j} & \boldsymbol{k} \\ -1 & 1 & 0 \\ 0 & -1 & 1 \end{vmatrix} = \boldsymbol{i} + \boldsymbol{j} + \boldsymbol{k}.$$

And the equation for the plane can be found from $\boldsymbol{N} \cdot \boldsymbol{r} = \boldsymbol{N} \cdot \boldsymbol{r}_1$, or

$$(\boldsymbol{i} + \boldsymbol{j} + \boldsymbol{k}) \cdot (x\boldsymbol{i} + y\boldsymbol{j} + z\boldsymbol{k}) = (\boldsymbol{i} + \boldsymbol{j} + \boldsymbol{k}) \cdot (2\boldsymbol{i} + \boldsymbol{j} + \boldsymbol{k}), \quad \text{or} \quad x + y + z = 4.$$

The intersection of this plane with the $z = 0$ plane forms the line given by $y = -x + 4$.

Problems for Lecture 6

1. Find an equation for the plane defined by the three points $(-1,-1,-1)$, $(1,1,1)$ and $(1,-1,0)$. Determine the equation for the line in the x-y plane formed by the intersection of this plane with the $z = 0$ plane.

Practice Quiz | Analytic geometry

1. The line that passes through the points $(0,1,1)$ and $(1,0,-1)$ has parametric equation given by

a) $t\boldsymbol{i} + (1+t)\boldsymbol{j} + (1+2t)\boldsymbol{k}$

b) $t\boldsymbol{i} + (1-t)\boldsymbol{j} + (1+2t)\boldsymbol{k}$

c) $t\boldsymbol{i} + (1+t)\boldsymbol{j} + (1-2t)\boldsymbol{k}$

d) $t\boldsymbol{i} + (1-t)\boldsymbol{j} + (1-2t)\boldsymbol{k}$

2. The line of Question 1 intersects the $z = 0$ plane at the point

a) $(\frac{1}{2}, \frac{1}{2}, 0)$

b) $(-\frac{1}{2}, \frac{1}{2}, 0)$

c) $(\frac{1}{2}, -\frac{1}{2}, 0)$

d) $(-\frac{1}{2}, -\frac{1}{2}, 0)$

3. The equation for the line in the x-y plane formed by the intersection of the plane defined by the points $(1,1,1)$, $(1,1,2)$ and $(2,1,1)$ and the $z = 0$ plane is given by

a) $y = x$

b) $y = x + 1$

c) $y = x - 1$

d) $y = 1$

Lecture 7 Kronecker delta and Levi-Civita symbol

We will soon make use of the Kronecker delta and Levi-Civita symbol to derive some important vector identities. We define the Kronecker delta δ_{ij} to be $+1$ if $i = j$ and 0 otherwise, and the Levi-Civita symbol ϵ_{ijk} to be $+1$ if i, j, and k are a cyclic permutation of $(1,2,3)$ (that is, one of $(1,2,3)$, $(2,3,1)$ or $(3,1,2)$); -1 if an anticyclic permutation of $(1,2,3)$ (that is, one of $(3,2,1)$, $(2,1,3)$ or $(1,3,2)$); and 0 if any two indices are equal. The choice of letters for the indices is arbitrary.

More formally, for indices restricted to the values of one, two or three,

$$\delta_{ij} = \begin{cases} 1, & \text{if } i = j; \\ 0, & \text{if } i \neq j; \end{cases}$$

and

$$\epsilon_{ijk} = \begin{cases} +1, & \text{if } (i,j,k) \text{ is } (1,2,3), (2,3,1) \text{ or } (3,1,2); \\ -1, & \text{if } (i,j,k) \text{ is } (3,2,1), (2,1,3) \text{ or } (1,3,2); \\ 0, & \text{if } i = j \text{ or } j = k \text{ or } k = i. \end{cases}$$

For convenience, we will use the Einstein summation convention when working with these symbols, where a repeated index within a single term or in a product of terms implies summation over that index. For example, we will write an expression such as $\sum_{i=1}^{3} A_i B_i$ more simply as $A_i B_i$. Within a single term or a product of terms, any index letter can be repeated at most one time, though multiple index letters may be repeated.

Using the Einstein summation convention, some interesting examples are

$$\delta_{ii} = 3 \quad \text{and} \quad \epsilon_{ijk}\epsilon_{ijk} = 6,$$

where $\delta_{ii} = \delta_{11} + \delta_{22} + \delta_{33}$ and $\epsilon_{ijk}\epsilon_{ijk}$ implies a sum over i, j, and k, containing a total of $3^3 = 27$ terms in the sum, where six of the terms are equal to one and the remaining terms are equal to zero. As these two examples show, repeated indices in a term or product may dramatically simplify the expression.

Finally, we state here a remarkable relationship between the product of Levi-Civita symbols and the Kronecker delta, given by the determinant

$$\begin{aligned} \epsilon_{ijk}\epsilon_{lmn} &= \begin{vmatrix} \delta_{il} & \delta_{im} & \delta_{in} \\ \delta_{jl} & \delta_{jm} & \delta_{jn} \\ \delta_{kl} & \delta_{km} & \delta_{kn} \end{vmatrix} \\ &= \delta_{il}(\delta_{jm}\delta_{kn} - \delta_{jn}\delta_{km}) - \delta_{im}(\delta_{jl}\delta_{kn} - \delta_{jn}\delta_{kl}) + \delta_{in}(\delta_{jl}\delta_{km} - \delta_{jm}\delta_{kl}). \end{aligned}$$

Problems for Lecture 7

1. Prove the following cyclic and anticyclic permutation identities:

a) $\epsilon_{ijk} = \epsilon_{jki} = \epsilon_{kij}$;

b) $\epsilon_{ijk} = -\epsilon_{jik}, \quad \epsilon_{ijk} = -\epsilon_{kji}, \quad \epsilon_{ijk} = -\epsilon_{ikj}$.

2. The notation $[\dots]_i$ means the ith component of the bracketed vector. Verify the cross-product relation $[\boldsymbol{A} \times \boldsymbol{B}]_i = \epsilon_{ijk} A_j B_k$ by considering $i = 1, 2, 3$.

3. Prove the following Kronecker-delta identities:

a) $\delta_{ij} A_j = A_i$;

b) $\delta_{ik} \delta_{kj} = \delta_{ij}$.

4. Given the most general identity relating the Levi-Civita symbol to the Kronecker delta,

$$\epsilon_{ijk}\epsilon_{lmn} = \delta_{il}(\delta_{jm}\delta_{kn} - \delta_{jn}\delta_{km}) - \delta_{im}(\delta_{jl}\delta_{kn} - \delta_{jn}\delta_{kl}) + \delta_{in}(\delta_{jl}\delta_{km} - \delta_{jm}\delta_{kl}),$$

prove the following simpler and more useful relations:

a) $\epsilon_{ijk}\epsilon_{imn} = \delta_{jm}\delta_{kn} - \delta_{jn}\delta_{km}$;

b) $\epsilon_{ijk}\epsilon_{ijn} = 2\delta_{kn}$.

Lecture 8 | Vector identities

Four sometimes useful vector identities are

$$\boldsymbol{A} \cdot (\boldsymbol{B} \times \boldsymbol{C}) = \boldsymbol{B} \cdot (\boldsymbol{C} \times \boldsymbol{A}) = \boldsymbol{C} \cdot (\boldsymbol{A} \times \boldsymbol{B}), \qquad \text{(scalar triple product)}$$

$$\boldsymbol{A} \times (\boldsymbol{B} \times \boldsymbol{C}) = (\boldsymbol{A} \cdot \boldsymbol{C})\boldsymbol{B} - (\boldsymbol{A} \cdot \boldsymbol{B})\boldsymbol{C}, \qquad \text{(vector triple product)}$$

$$\begin{aligned}(\boldsymbol{A} \times \boldsymbol{B}) \cdot (\boldsymbol{C} \times \boldsymbol{D}) &= (\boldsymbol{A} \cdot \boldsymbol{C})(\boldsymbol{B} \cdot \boldsymbol{D}) \\ &\quad - (\boldsymbol{A} \cdot \boldsymbol{D})(\boldsymbol{B} \cdot \boldsymbol{C}), \qquad \text{(scalar quadruple product)}\end{aligned}$$

$$\begin{aligned}(\boldsymbol{A} \times \boldsymbol{B}) \times (\boldsymbol{C} \times \boldsymbol{D}) &= ((\boldsymbol{A} \times \boldsymbol{B}) \cdot \boldsymbol{D})\boldsymbol{C} \\ &\quad - ((\boldsymbol{A} \times \boldsymbol{B}) \cdot \boldsymbol{C})\boldsymbol{D}. \qquad \text{(vector quadruple product)}\end{aligned}$$

Parentheses are optional when expressions have only one possible interpretation, but for clarity they are often written. The first and third identities equate scalars whereas the second and fourth identities equate vectors. To prove two vectors $\boldsymbol{A}$ and $\boldsymbol{B}$ are equal, one must prove that their components are equal, that is $A_i = B_i$ for an arbitrary index i.

Proofs of these vector identities can make use of the Kronecker delta, the Levi-Civita symbol and the Einstein summation convention, where repeated indices are summed over. The algebraic toolbox that we will need contains the cyclic permutation identities for the Levi-Civita symbol:

$$\epsilon_{ijk} = \epsilon_{jki} = \epsilon_{kij};$$

the contraction of two Levi-Civita symbols:

$$\epsilon_{ijk}\epsilon_{imn} = \delta_{jm}\delta_{kn} - \delta_{jn}\delta_{km};$$

the contraction of the Kronecker delta with a vector:

$$\delta_{ij}A_j = A_i;$$

and the scalar and vector products written as

$$\boldsymbol{A} \cdot \boldsymbol{B} = A_i B_i, \qquad [\boldsymbol{A} \times \boldsymbol{B}]_i = \epsilon_{ijk} A_j B_k.$$

We will use the notation $[\dots]_i$ to mean the ith component of the bracketed vector.

Problems for Lecture 8

1. Remove all optional parentheses from the four vector identities:

$$\boldsymbol{A} \cdot (\boldsymbol{B} \times \boldsymbol{C}) = \boldsymbol{B} \cdot (\boldsymbol{C} \times \boldsymbol{A}) = \boldsymbol{C} \cdot (\boldsymbol{A} \times \boldsymbol{B}), \qquad \text{(scalar triple product)}$$

$$\boldsymbol{A} \times (\boldsymbol{B} \times \boldsymbol{C}) = (\boldsymbol{A} \cdot \boldsymbol{C})\boldsymbol{B} - (\boldsymbol{A} \cdot \boldsymbol{B})\boldsymbol{C}, \qquad \text{(vector triple product)}$$

$$(\boldsymbol{A} \times \boldsymbol{B}) \cdot (\boldsymbol{C} \times \boldsymbol{D}) = (\boldsymbol{A} \cdot \boldsymbol{C})(\boldsymbol{B} \cdot \boldsymbol{D}) - (\boldsymbol{A} \cdot \boldsymbol{D})(\boldsymbol{B} \cdot \boldsymbol{C}), \qquad \text{(scalar quadruple product)}$$

$$(\boldsymbol{A} \times \boldsymbol{B}) \times (\boldsymbol{C} \times \boldsymbol{D}) = ((\boldsymbol{A} \times \boldsymbol{B}) \cdot \boldsymbol{D})\boldsymbol{C} - ((\boldsymbol{A} \times \boldsymbol{B}) \cdot \boldsymbol{C})\boldsymbol{D}. \qquad \text{(vector quadruple product)}$$

Lecture 9 | Scalar triple product

The scalar triple product, which can be written as either $\boldsymbol{A} \cdot (\boldsymbol{B} \times \boldsymbol{C})$ or $\boldsymbol{A} \cdot \boldsymbol{B} \times \boldsymbol{C}$, satisfies

$$\boldsymbol{A} \cdot (\boldsymbol{B} \times \boldsymbol{C}) = \boldsymbol{B} \cdot (\boldsymbol{C} \times \boldsymbol{A}) = \boldsymbol{C} \cdot (\boldsymbol{A} \times \boldsymbol{B}),$$

that is, its value is unchanged under a cyclic permutation of the three vectors. This identity can be proved using the cyclic property of the Levi-Civita symbol:

$$\boldsymbol{A} \cdot (\boldsymbol{B} \times \boldsymbol{C}) = A_i \epsilon_{ijk} B_j C_k = B_j \epsilon_{jki} C_k A_i = \boldsymbol{B} \cdot (\boldsymbol{C} \times \boldsymbol{A}),$$

and similarly,

$$\boldsymbol{B} \cdot (\boldsymbol{C} \times \boldsymbol{A}) = B_j \epsilon_{jki} C_k A_i = C_k \epsilon_{kij} A_i B_j = \boldsymbol{C} \cdot (\boldsymbol{A} \times \boldsymbol{B}).$$

We can also write the scalar triple product as a three-by-three determinant. Using the determinant expression for the cross-product, we have

$$\boldsymbol{A} \cdot (\boldsymbol{B} \times \boldsymbol{C}) = \boldsymbol{A} \cdot \begin{vmatrix} \boldsymbol{i} & \boldsymbol{j} & \boldsymbol{k} \\ B_1 & B_2 & B_3 \\ C_1 & C_2 & C_3 \end{vmatrix} = \begin{vmatrix} A_1 & A_2 & A_3 \\ B_1 & B_2 & B_3 \\ C_1 & C_2 & C_3 \end{vmatrix}.$$

The absolute value of a three-by-three determinant, and therefore that of the scalar triple product, is the volume of the parallelepiped defined by the three row vectors $\boldsymbol{A}$, $\boldsymbol{B}$ and $\boldsymbol{C}$, as shown in the figure below.

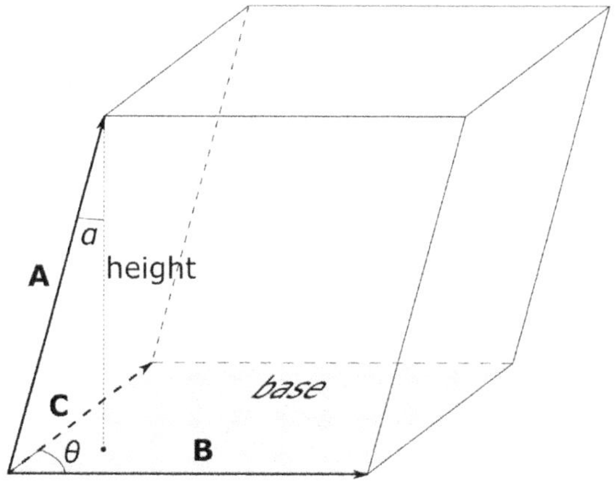

Problems for Lecture 9

1. Consider the scalar triple product $\boldsymbol{A} \cdot (\boldsymbol{B} \times \boldsymbol{C})$. Prove that if any two vectors are equal, then the scalar triple product is zero.

2. Show that swapping the positions of the operators without re-ordering the vectors leaves the scalar triple product unchanged, that is,

$$\boldsymbol{A} \cdot \boldsymbol{B} \times \boldsymbol{C} = \boldsymbol{A} \times \boldsymbol{B} \cdot \boldsymbol{C}.$$

3. It is sometimes useful to define a notation where the unit vectors are distinguished by their index. That is, we define $\boldsymbol{e}_1 = \boldsymbol{i}$, $\boldsymbol{e}_2 = \boldsymbol{j}$ and $\boldsymbol{e}_3 = \boldsymbol{k}$. Prove that

$$\boldsymbol{e}_i \cdot (\boldsymbol{e}_j \times \boldsymbol{e}_k) = \epsilon_{ijk},$$

where ϵ_{ijk} is the usual Levi-Civita symbol.

Lecture 10 Vector triple product

The vector triple product satisfies

$$\boldsymbol{A} \times (\boldsymbol{B} \times \boldsymbol{C}) = (\boldsymbol{A} \cdot \boldsymbol{C})\boldsymbol{B} - (\boldsymbol{A} \cdot \boldsymbol{B})\boldsymbol{C}.$$

This identity can be proved using Kronecker delta and Levi-Civita symbol identities as follows:

$$\begin{aligned}
& & & \underline{\text{Justification}} \\
[\boldsymbol{A} \times (\boldsymbol{B} \times \boldsymbol{C})]_i &= \epsilon_{ijk} A_j [\boldsymbol{B} \times \boldsymbol{C}]_k & & [\boldsymbol{A} \times \boldsymbol{X}]_i = \epsilon_{ijk} A_j X_k \\
&= \epsilon_{ijk} A_j \epsilon_{klm} B_l C_m & & [\boldsymbol{B} \times \boldsymbol{C}]_k = \epsilon_{klm} B_l C_m \\
&= \epsilon_{kij} \epsilon_{klm} A_j B_l C_m & & \epsilon_{ijk} = \epsilon_{kij} \\
&= (\delta_{il}\delta_{jm} - \delta_{im}\delta_{jl}) A_j B_l C_m & & \epsilon_{kij}\epsilon_{klm} = \delta_{il}\delta_{jm} - \delta_{im}\delta_{jl} \\
&= A_j B_i C_j - A_j B_j C_i & & \delta_{il} B_l = B_i,\ \delta_{jm} C_m = C_j, \text{ etc.} \\
&= (\boldsymbol{A} \cdot \boldsymbol{C}) B_i - (\boldsymbol{A} \cdot \boldsymbol{B}) C_i & & A_j C_j = \boldsymbol{A} \cdot \boldsymbol{C},\ A_j B_j = \boldsymbol{A} \cdot \boldsymbol{B} \\
&= [(\boldsymbol{A} \cdot \boldsymbol{C})\boldsymbol{B} - (\boldsymbol{A} \cdot \boldsymbol{B})\boldsymbol{C}]_i. & & \text{vector subtraction law}
\end{aligned}$$

Problems for Lecture 10

1. Prove the Jacobi identity:

$$\boldsymbol{A} \times (\boldsymbol{B} \times \boldsymbol{C}) + \boldsymbol{B} \times (\boldsymbol{C} \times \boldsymbol{A}) + \boldsymbol{C} \times (\boldsymbol{A} \times \boldsymbol{B}) = 0.$$

2. Prove that the scalar quadruple product satisfies

$$(\boldsymbol{A} \times \boldsymbol{B}) \cdot (\boldsymbol{C} \times \boldsymbol{D}) = (\boldsymbol{A} \cdot \boldsymbol{C})(\boldsymbol{B} \cdot \boldsymbol{D}) - (\boldsymbol{A} \cdot \boldsymbol{D})(\boldsymbol{B} \cdot \boldsymbol{C}).$$

3. Prove Lagrange's identity in three dimensions:

$$|\boldsymbol{A} \times \boldsymbol{B}|^2 = |\boldsymbol{A}|^2|\boldsymbol{B}|^2 - (\boldsymbol{A} \cdot \boldsymbol{B})^2.$$

4. Prove using vector triple products that the vector quadruple product satisfies

$$(\boldsymbol{A} \times \boldsymbol{B}) \times (\boldsymbol{C} \times \boldsymbol{D}) = ((\boldsymbol{A} \times \boldsymbol{B}) \cdot \boldsymbol{D})\boldsymbol{C} - ((\boldsymbol{A} \times \boldsymbol{B}) \cdot \boldsymbol{C})\boldsymbol{D}.$$

Practice Quiz | Vector algebra

1. The expression $\epsilon_{ijk}\epsilon_{ljm}$ is equal to

    a) $\delta_{jm}\delta_{kn} - \delta_{jn}\delta_{km}$

    b) $\delta_{jn}\delta_{km} - \delta_{jm}\delta_{kn}$

    c) $\delta_{km}\delta_{il} - \delta_{kl}\delta_{im}$

    d) $\delta_{kl}\delta_{im} - \delta_{km}\delta_{il}$

2. The expression $\boldsymbol{A} \times (\boldsymbol{B} \times \boldsymbol{C})$ is always equal to

    a) $\boldsymbol{B} \times (\boldsymbol{C} \times \boldsymbol{A})$

    b) $\boldsymbol{A} \times (\boldsymbol{C} \times \boldsymbol{B})$

    c) $(\boldsymbol{A} \times \boldsymbol{B}) \times \boldsymbol{C}$

    d) $(\boldsymbol{C} \times \boldsymbol{B}) \times \boldsymbol{A}$

3. Which of the following expressions may not be zero?

    a) $\boldsymbol{A} \cdot (\boldsymbol{B} \times \boldsymbol{B})$

    b) $\boldsymbol{A} \cdot (\boldsymbol{A} \times \boldsymbol{B})$

    c) $\boldsymbol{A} \times (\boldsymbol{A} \times \boldsymbol{B})$

    d) $\boldsymbol{B} \cdot (\boldsymbol{A} \times \boldsymbol{B})$

Lecture 11 Scalar and vector fields

In physics, scalars and vectors can be functions of both space and time. We call these types of functions fields. For example, the temperature in some region of space is a scalar field, and we can write

$$T(\boldsymbol{r}, t) = T(x, y, z; t),$$

where we use the common notation of a semicolon on the right-hand side to separate the space and time dependence. Notice that the position vector $\boldsymbol{r}$ is used to locate the temperature in space. As another example, the velocity vector $\boldsymbol{u}$ of a flowing fluid is a vector field, and if the components of this velocity field are u_1, u_2 and u_3, we can write

$$\boldsymbol{u}(\boldsymbol{r}, t) = u_1(x, y, z; t)\boldsymbol{i} + u_2(x, y, z; t)\boldsymbol{j} + u_3(x, y, z; t)\boldsymbol{k}.$$

The equations governing a field are called the field equations, and these equations commonly take the form of partial differential equations. For example, the equations for the electric and magnetic vector fields are the famous Maxwell's equations, and the equation for the fluid velocity vector field is called the Navier-Stokes equation. The equation for the scalar field (called the wave function) in non-relativistic quantum mechanics is called the Schrödinger equation.

There are many ways to visualize scalar and vector fields, and one tries to make plots as informative as possible. On the right is a simple visualization of the two-dimensional vector field given by

$$\boldsymbol{B}(x, y) = \frac{-y\,\boldsymbol{i} + x\,\boldsymbol{j}}{x^2 + y^2},$$

where the vectors at each point are represented by arrows.

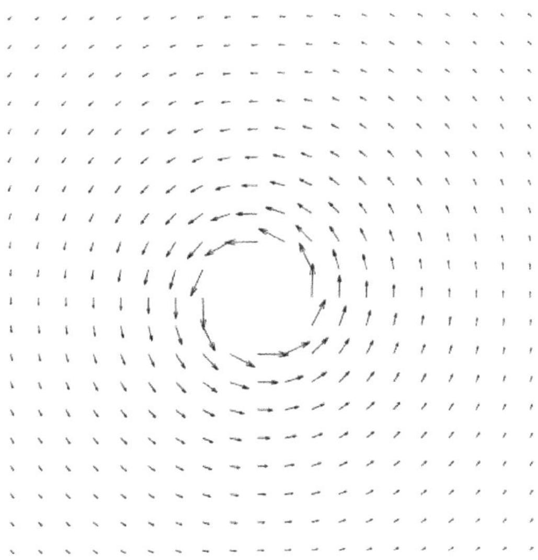

Problems for Lecture 11

1. List some physical examples of scalar and vector fields.

Week II

Differentiation

In this week's lectures, we learn about the derivatives of scalar and vector fields. We define the partial derivative and derive the method of least squares as a minimization problem. We learn how to use the chain rule for a function of several variables, and derive the triple product rule used in chemistry. From the del differential operator, we define the gradient, divergence, curl and Laplacian. We learn some useful vector derivative identities and how to derive them using the Kronecker delta and Levi-Civita symbol. Vector identities are then used to derive the electromagnetic wave equation from Maxwell's equations in free space. Electromagnetic waves are fundamental to all modern communication technologies.

Lecture 12 Partial derivatives

For a function $f = f(x, y)$ of two variables, we define the partial derivative of f with respect to x as

$$\frac{\partial f}{\partial x} = \lim_{h \to 0} \frac{f(x+h, y) - f(x, y)}{h},$$

and similarly for the partial derivative of f with respect to y. To take a partial derivative with respect to a variable, take the derivative with respect to that variable treating all other variables as constants. As an example, consider

$$f(x, y) = 2x^3y^2 + y^3.$$

We have

$$\frac{\partial f}{\partial x} = 6x^2y^2, \quad \frac{\partial f}{\partial y} = 4x^3y + 3y^2.$$

Second derivatives are defined as the derivatives of the first derivatives, so we have

$$\frac{\partial^2 f}{\partial x^2} = 12xy^2, \quad \frac{\partial^2 f}{\partial y^2} = 4x^3 + 6y;$$

and for continuous differentiable functions, the mixed second partial derivatives are independent of the order in which the derivatives are taken,

$$\frac{\partial^2 f}{\partial x \partial y} = 12x^2y = \frac{\partial^2 f}{\partial y \partial x}.$$

To simplify notation, we introduce the standard subscript notation for partial derivatives,

$$f_x = \frac{\partial f}{\partial x}, \quad f_y = \frac{\partial f}{\partial y}, \quad f_{xx} = \frac{\partial^2 f}{\partial x^2}, \quad f_{xy} = \frac{\partial^2 f}{\partial x \partial y}, \quad f_{yy} = \frac{\partial^2 f}{\partial y^2}, \quad \text{etc.}$$

The Taylor series of $f(x, y)$ about the origin is developed by expanding the function in a multivariable power series that agrees with the function value and all its partial derivatives at $(x, y) = (0, 0)$. We have

$$f(x, y) = f + f_x x + f_y y + \frac{1}{2!}\left(f_{xx}x^2 + 2f_{xy}xy + f_{yy}y^2\right) + \ldots.$$

The function and all its partial derivatives on the right-hand side are evaluated at $(0, 0)$ and are constants. By taking partial derivatives it is evident that $f(0, 0) = f$, $f_x(0, 0) = f_x$, $f_y(0, 0) = f_y$, and so on as the infinite series continues.

Problems for Lecture 12

1. Compute the three partial derivatives of

$$f(x, y, z) = \frac{1}{(x^2 + y^2 + z^2)^n}.$$

2. Given the function $f = f(t, x)$, find the Taylor series expansion of the expression

$$f(t + \alpha\Delta t, x + \beta\Delta t f(t, x))$$

to first order in Δt.

Lecture 13 The method of least squares

Local maxima and minima of a multivariable function can be found by computing the zeros of the partial derivatives. These zeros are called critical points. A critical point need not be a maximum or minimum, for example it might be a minimum in one direction and a maximum in another (called a saddle point), but in many problems these points are maxima or minima. Here, we will solve the least-squares problem by minimizing a function.

Suppose there is some experimental data that you want to fit by a straight line (illustrated on the right). In general, let the data consist of a set of n points given by (x_1, y_1), ..., (x_n, y_n). Here, we assume that the x values are exact, and the y values are noisy. We further assume that the best fit line to the data takes the form $y = \beta_0 + \beta_1 x$. Although we know that the line can not go through all the data points, we can try to find the line that minimizes the sum of the squares of the vertical distances between the line and the points.

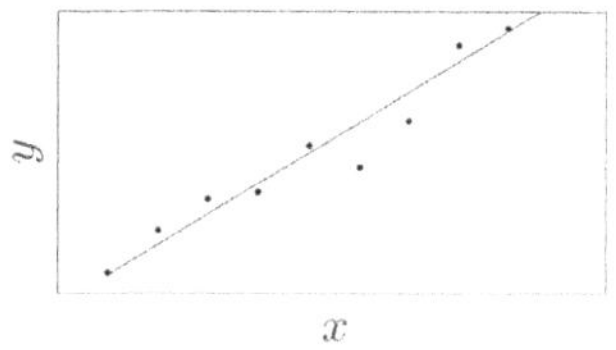

Define this function of the sum of the squares to be

$$f(\beta_0, \beta_1) = \sum_{i=1}^{n} (\beta_0 + \beta_1 x_i - y_i)^2 .$$

Here, the data are assumed given and the unknowns are the fitting parameters β_0 and β_1. It should be clear from the problem specification, that there must be values of β_0 and β_1 that minimize the function $f = f(\beta_0, \beta_1)$. To determine, these values, we set $\partial f/\partial \beta_0 = \partial f/\partial \beta_1 = 0$. This results in the equations

$$\sum_{i=1}^{n}(\beta_0 + \beta_1 x_i - y_i) = 0, \qquad \sum_{i=1}^{n} x_i (\beta_0 + \beta_1 x_i - y_i) = 0.$$

We can write these equations as a linear system for β_0 and β_1 as

$$\beta_0 n + \beta_1 \sum_{i=1}^{n} x_i = \sum_{i=1}^{n} y_i , \qquad \beta_0 \sum_{i=1}^{n} x_i + \beta_1 \sum_{i=1}^{n} x_i^2 = \sum_{i=1}^{n} x_i y_i .$$

The solution for β_0 and β_1 in terms of the data is given by

$$\beta_0 = \frac{\sum x_i^2 \sum y_i - \sum x_i y_i \sum x_i}{n \sum x_i^2 - (\sum x_i)^2}, \quad \beta_1 = \frac{n \sum x_i y_i - (\sum x_i)(\sum y_i)}{n \sum x_i^2 - (\sum x_i)^2},$$

where the summations are from $i = 1$ to n.

Problems for Lecture 13

1. By minimizing the sum of the squares of the vertical distance between the line and the points, determine the least-squares line through the data points $(1, 1)$, $(2, 3)$ and $(3, 2)$.

Lecture 14 | Chain rule

Partial derivatives are used in applying the chain rule to a function of several variables. Consider a two-dimensional scalar field $f = f(x,y)$, and define the total differential of f to be

$$df = f(x + dx, y + dy) - f(x,y).$$

We can write df as

$$\begin{aligned} df &= [f(x + dx, y + dy) - f(x, y + dy)] + [f(x, y + dy) - f(x,y)] \\ &= \frac{\partial f}{\partial x}dx + \frac{\partial f}{\partial y}dy. \end{aligned}$$

If one has $f = f(x(t), y(t))$, say, then division of df by dt results in

$$\frac{df}{dt} = \frac{\partial f}{\partial x}\frac{dx}{dt} + \frac{\partial f}{\partial y}\frac{dy}{dt}.$$

And if one has $f = f(x(r,\theta), y(r,\theta))$, say, then the corresponding chain rule is given by

$$\frac{\partial f}{\partial r} = \frac{\partial f}{\partial x}\frac{\partial x}{\partial r} + \frac{\partial f}{\partial y}\frac{\partial y}{\partial r}, \qquad \frac{\partial f}{\partial \theta} = \frac{\partial f}{\partial x}\frac{\partial x}{\partial \theta} + \frac{\partial f}{\partial y}\frac{\partial y}{\partial \theta}.$$

Example: Consider the differential equation $\frac{dx}{dt} = u(t, x(t))$. *Determine a formula for* $\frac{d^2x}{dt^2}$ *in terms of u and its partial derivatives.*

Applying the chain rule, we have at time t,

$$\begin{aligned} \frac{d^2x}{dt^2} &= \frac{\partial u}{\partial t} + \frac{\partial u}{\partial x}\frac{dx}{dt} \\ &= \frac{\partial u}{\partial t} + u\frac{\partial u}{\partial x}. \end{aligned}$$

The above formula is called the material derivative and in three dimensions forms a part of the Navier-Stokes equation for fluid flow.

Problems for Lecture 14

1. Let $f(x,y) = e^{xy}$, with $x = r\cos\theta$ and $y = r\sin\theta$. Compute the partial derivatives $\partial f/\partial r$ and $\partial f/\partial \theta$ in two ways:

a) Use the chain rule on $f = f(x(r,\theta), y(r,\theta))$;

b) Eliminate x and y in favor of r and θ and compute the partial derivatives directly.

Lecture 15 Triple product rule

Suppose that three variables x, y and z are related by the equation $f(x,y,z) = 0$, and that it is possible to write $x = x(y,z)$, $y = y(x,z)$ and $z = z(x,y)$. Taking differentials of x and z, we have

$$dx = \frac{\partial x}{\partial y}dy + \frac{\partial x}{\partial z}dz, \qquad dz = \frac{\partial z}{\partial x}dx + \frac{\partial z}{\partial y}dy.$$

We can make use of the second equation to eliminate dz in the first equation to obtain

$$dx = \frac{\partial x}{\partial y}dy + \frac{\partial x}{\partial z}\left(\frac{\partial z}{\partial x}dx + \frac{\partial z}{\partial y}dy\right);$$

or collecting terms,

$$\left(1 - \frac{\partial x}{\partial z}\frac{\partial z}{\partial x}\right)dx = \left(\frac{\partial x}{\partial y} + \frac{\partial x}{\partial z}\frac{\partial z}{\partial y}\right)dy.$$

Since dx and dy can be independent variations, the terms in parentheses must be zero. The left-hand-side results in the reciprocity relation

$$\frac{\partial x}{\partial z}\frac{\partial z}{\partial x} = 1,$$

which states the intuitive result that $\partial z/\partial x$ and $\partial x/\partial z$ are multiplicative inverses of each other. The right-hand-side results in

$$\frac{\partial x}{\partial y} = -\frac{\partial x}{\partial z}\frac{\partial z}{\partial y},$$

which when making use of the reciprocity relation, yields the counterintuitive triple product rule,

$$\frac{\partial x}{\partial y}\frac{\partial y}{\partial z}\frac{\partial z}{\partial x} = -1.$$

Lecture 16 Triple product rule (example)

Example: Demonstrate the triple product rule using the ideal gas law.

The ideal gas law states that

$$PV = nRT,$$

where P is the pressure, V is the volume, T is the absolute temperature, n is the number of moles of the gas, and R is the ideal gas constant. We say P, V and T are the state variables, and the ideal gas law is a relation of the form

$$f(P, V, T) = PV - nRT = 0.$$

We can write $P = P(V, T)$, $V = V(P, T)$ and $T = T(P, V)$, that is,

$$P = \frac{nRT}{V}, \qquad V = \frac{nRT}{P}, \qquad T = \frac{PV}{nR};$$

and the partial derivatives are given by

$$\frac{\partial P}{\partial V} = -\frac{nRT}{V^2}, \qquad \frac{\partial V}{\partial T} = \frac{nR}{P}, \qquad \frac{\partial T}{\partial P} = \frac{V}{nR}.$$

The triple product results in

$$\frac{\partial P}{\partial V}\frac{\partial V}{\partial T}\frac{\partial T}{\partial P} = -\left(\frac{nRT}{V^2}\right)\left(\frac{nR}{P}\right)\left(\frac{V}{nR}\right) = -\frac{nRT}{PV} = -1,$$

where we make use of the ideal gas law in the last equality.

Problems for Lecture 16

1. Suppose the three variables x, y and z are related by the linear expression $ax + by + cz = 0$. Show that x, y and z satisfy the triple product rule.

2. Suppose the four variables x, y, z and t are related by the linear expression $ax + by + cz + dt = 0$. Determine a corresponding quadruple product rule for these variables.

Practice Quiz | Partial derivatives

1. Let $f(x,y,z) = \dfrac{1}{(x^2+y^2+z^2)^{1/2}}$. The mixed second partial derivative $\dfrac{\partial^2 f}{\partial x \partial y}$ is equal to

a) $\dfrac{2(x+y)}{(x^2+y^2+z^2)^{5/2}}$

b) $\dfrac{(x+y)^2}{(x^2+y^2+z^2)^{5/2}}$

c) $\dfrac{x^2+y^2}{(x^2+y^2+z^2)^{5/2}}$

d) $\dfrac{3xy}{(x^2+y^2+z^2)^{5/2}}$

2. The least-squares line through the data points $(0,1)$, $(1,3)$, $(2,3)$ and $(3,4)$ is given by

a) $y = \dfrac{7}{5} + \dfrac{9x}{10}$

b) $y = \dfrac{5}{7} + \dfrac{9x}{10}$

c) $y = \dfrac{7}{5} + \dfrac{10x}{9}$

d) $y = \dfrac{5}{7} + \dfrac{10x}{9}$

3. Let $f = f(x,y)$ with $x = r\cos\theta$ and $y = r\sin\theta$. Which of the following is true?

a) $\dfrac{\partial f}{\partial \theta} = x\dfrac{\partial f}{\partial x} + y\dfrac{\partial f}{\partial y}$

b) $\dfrac{\partial f}{\partial \theta} = -x\dfrac{\partial f}{\partial x} + y\dfrac{\partial f}{\partial y}$

c) $\dfrac{\partial f}{\partial \theta} = y\dfrac{\partial f}{\partial x} + x\dfrac{\partial f}{\partial y}$

d) $\dfrac{\partial f}{\partial \theta} = -y\dfrac{\partial f}{\partial x} + x\dfrac{\partial f}{\partial y}$

Lecture 17 | Gradient

Consider the three-dimensional scalar field $f = f(x, y, z)$, and the differential df, given by

$$df = \frac{\partial f}{\partial x}dx + \frac{\partial f}{\partial y}dy + \frac{\partial f}{\partial z}dz.$$

Using the dot product, we can write this in vector form as

$$df = \left(\frac{\partial f}{\partial x}\boldsymbol{i} + \frac{\partial f}{\partial y}\boldsymbol{j} + \frac{\partial f}{\partial z}\boldsymbol{k}\right) \cdot (dx\boldsymbol{i} + dy\boldsymbol{j} + dz\boldsymbol{k}) = \boldsymbol{\nabla} f \cdot d\boldsymbol{r},$$

where $d\boldsymbol{r} = dx\boldsymbol{i} + dy\boldsymbol{j} + dz\boldsymbol{k}$, and

$$\boldsymbol{\nabla} f = \frac{\partial f}{\partial x}\boldsymbol{i} + \frac{\partial f}{\partial y}\boldsymbol{j} + \frac{\partial f}{\partial z}\boldsymbol{k}$$

is called the gradient of f. The nabla symbol $\boldsymbol{\nabla}$ is pronounced "del" and $\boldsymbol{\nabla} f$ is pronounced "del-f". Another useful way to view the gradient is to consider $\boldsymbol{\nabla}$ as a vector differential operator which has the form

$$\boldsymbol{\nabla} = \boldsymbol{i}\frac{\partial}{\partial x} + \boldsymbol{j}\frac{\partial}{\partial y} + \boldsymbol{k}\frac{\partial}{\partial z}.$$

Because of the properties of the dot product, the differential df is maximum when the infinitesimal displacement vector $d\boldsymbol{r}$ points in the same direction as the gradient vector $\boldsymbol{\nabla} f$. Therefore, $\boldsymbol{\nabla} f$ points in the direction of maximally increasing f, and the magnitude of $\boldsymbol{\nabla} f$ gives the slope (or gradient) of f in that direction.

Example: Compute the gradient of $f(x, y, z) = xyz$.
The partial derivatives are easily calculated, and we have

$$\boldsymbol{\nabla} f = yz\,\boldsymbol{i} + xz\,\boldsymbol{j} + xy\,\boldsymbol{k}.$$

Problems for Lecture 17

1. Let the position vector be given by $\boldsymbol{r} = x\boldsymbol{i} + y\boldsymbol{j} + z\boldsymbol{k}$ and its length be given by $r = \sqrt{x^2 + y^2 + z^2}$. Compute the gradient of the following scalar fields and write your results in terms of $\boldsymbol{r}$ and r.

a) $\phi(x, y, z) = x^2 + y^2 + z^2$;

b) $\phi(x, y, z) = \sqrt{x^2 + y^2 + z^2}$;

c) $\phi(x, y, z) = \dfrac{1}{\sqrt{x^2 + y^2 + z^2}}$.

2. Using the results from Problem 1, guess the general form of $\boldsymbol{\nabla}(r^n)$.

Lecture 18 Divergence

Consider in Cartesian coordinates the three-dimensional vector field, $\boldsymbol{u} = u_1(x,y,z)\boldsymbol{i} + u_2(x,y,z)\boldsymbol{j} + u_3(x,y,z)\boldsymbol{k}$. The divergence of $\boldsymbol{u}$, denoted as $\boldsymbol{\nabla} \cdot \boldsymbol{u}$ and pronounced "del-dot-u", is defined as the scalar field given by

$$\begin{aligned}\boldsymbol{\nabla} \cdot \boldsymbol{u} &= \left(\boldsymbol{i}\frac{\partial}{\partial x} + \boldsymbol{j}\frac{\partial}{\partial y} + \boldsymbol{k}\frac{\partial}{\partial z}\right) \cdot (u_1\boldsymbol{i} + u_2\boldsymbol{j} + u_3\boldsymbol{k}) \\ &= \frac{\partial u_1}{\partial x} + \frac{\partial u_2}{\partial y} + \frac{\partial u_3}{\partial z}.\end{aligned}$$

Here, the dot product is used between a vector differential operator $\boldsymbol{\nabla}$ and a vector field $\boldsymbol{u}$. The divergence measures how much a vector field spreads out, or diverges, from a point. A more math-based description will be given later.

Example: Let the position vector be given by $\boldsymbol{r} = x\boldsymbol{i} + y\boldsymbol{j} + z\boldsymbol{k}$. Find $\boldsymbol{\nabla} \cdot \boldsymbol{r}$.

A direct calculation gives

$$\boldsymbol{\nabla} \cdot \boldsymbol{r} = \frac{\partial}{\partial x}x + \frac{\partial}{\partial y}y + \frac{\partial}{\partial z}z = 3.$$

Example: Let $\boldsymbol{F} = \dfrac{\boldsymbol{r}}{|\boldsymbol{r}|^3}$ for all $\boldsymbol{r} \neq 0$. Find $\boldsymbol{\nabla} \cdot \boldsymbol{F}$.

Writing out the components of $\boldsymbol{F}$, we have

$$\begin{aligned}\boldsymbol{F} &= F_1\boldsymbol{i} + F_2\boldsymbol{j} + F_3\boldsymbol{k} \\ &= \frac{x}{(x^2+y^2+z^2)^{3/2}}\boldsymbol{i} + \frac{y}{(x^2+y^2+z^2)^{3/2}}\boldsymbol{j} + \frac{z}{(x^2+y^2+z^2)^{3/2}}\boldsymbol{k}.\end{aligned}$$

Using the quotient rule for the derivative, we have

$$\frac{\partial F_1}{\partial x} = \frac{(x^2+y^2+z^2)^{3/2} - 3x^2(x^2+y^2+z^2)^{1/2}}{(x^2+y^2+z^2)^3} = \frac{1}{|\boldsymbol{r}|^3} - \frac{3x^2}{|\boldsymbol{r}|^5},$$

and analogous results for $\partial F_2/\partial y$ and $\partial F_3/\partial z$. Adding the three derivatives results in

$$\boldsymbol{\nabla} \cdot \boldsymbol{F} = \frac{3}{|\boldsymbol{r}|^3} - \frac{3(x^2+y^2+z^2)}{|\boldsymbol{r}|^5} = \frac{3}{|\boldsymbol{r}|^3} - \frac{3}{|\boldsymbol{r}|^3} = 0,$$

valid as long as $|\boldsymbol{r}| \neq 0$, where $\boldsymbol{F}$ diverges. In electrostatics, $\boldsymbol{F}$ is proportional to the electric field of a point charge located at the origin.

Problems for Lecture 18

1. Find the divergence of the following vector fields:

a) $\boldsymbol{F} = xy\boldsymbol{i} + yz\boldsymbol{j} + zx\boldsymbol{k}$;

b) $\boldsymbol{F} = yz\boldsymbol{i} + xz\boldsymbol{j} + xy\boldsymbol{k}$.

Lecture 19 Curl

Consider in Cartesian coordinates the three-dimensional vector field $\boldsymbol{u} = u_1(x,y,z)\boldsymbol{i} + u_2(x,y,z)\boldsymbol{j} + u_3(x,y,z)\boldsymbol{k}$. The curl of $\boldsymbol{u}$, denoted as $\boldsymbol{\nabla} \times \boldsymbol{u}$ and pronounced "del-cross-u", is defined as the vector field given by

$$\begin{aligned}\boldsymbol{\nabla} \times \boldsymbol{u} &= \begin{vmatrix} \boldsymbol{i} & \boldsymbol{j} & \boldsymbol{k} \\ \partial/\partial x & \partial/\partial y & \partial/\partial z \\ u_1 & u_2 & u_3 \end{vmatrix} \\ &= \left(\frac{\partial u_3}{\partial y} - \frac{\partial u_2}{\partial z}\right)\boldsymbol{i} + \left(\frac{\partial u_1}{\partial z} - \frac{\partial u_3}{\partial x}\right)\boldsymbol{j} + \left(\frac{\partial u_2}{\partial x} - \frac{\partial u_1}{\partial y}\right)\boldsymbol{k}.\end{aligned}$$

Here, the cross product is used between a vector differential operator and a vector field. The curl measures how much a vector field rotates, or curls, around a point. A more math-based description will be given later.

Example: Show that the curl of a gradient is zero, that is, $\boldsymbol{\nabla} \times (\boldsymbol{\nabla} f) = 0$.
We have

$$\begin{aligned}\boldsymbol{\nabla} \times (\boldsymbol{\nabla} f) &= \begin{pmatrix} \boldsymbol{i} & \boldsymbol{j} & \boldsymbol{k} \\ \partial/\partial x & \partial/\partial y & \partial/\partial z \\ \partial f/\partial x & \partial f/\partial y & \partial f/\partial z \end{pmatrix} \\ &= \left(\frac{\partial^2 f}{\partial y \partial z} - \frac{\partial^2 f}{\partial z \partial y}\right)\boldsymbol{i} + \left(\frac{\partial^2 f}{\partial z \partial x} - \frac{\partial^2 f}{\partial x \partial z}\right)\boldsymbol{j} + \left(\frac{\partial^2 f}{\partial x \partial y} - \frac{\partial^2 f}{\partial y \partial x}\right)\boldsymbol{k} \\ &= 0,\end{aligned}$$

using the equality of mixed partials.

Example: Show that the divergence of a curl is zero, that is, $\boldsymbol{\nabla} \cdot (\boldsymbol{\nabla} \times \boldsymbol{u}) = 0$.
We have

$$\begin{aligned}\boldsymbol{\nabla} \cdot (\boldsymbol{\nabla} \times \boldsymbol{u}) &= \frac{\partial}{\partial x}\left(\frac{\partial u_3}{\partial y} - \frac{\partial u_2}{\partial z}\right) + \frac{\partial}{\partial y}\left(\frac{\partial u_1}{\partial z} - \frac{\partial u_3}{\partial x}\right) + \frac{\partial}{\partial z}\left(\frac{\partial u_2}{\partial x} - \frac{\partial u_1}{\partial y}\right) \\ &= \left(\frac{\partial^2 u_1}{\partial y \partial z} - \frac{\partial^2 u_1}{\partial z \partial y}\right) + \left(\frac{\partial^2 u_2}{\partial z \partial x} - \frac{\partial^2 u_2}{\partial x \partial z}\right) + \left(\frac{\partial^2 u_3}{\partial x \partial y} - \frac{\partial^2 u_3}{\partial y \partial x}\right) \\ &= 0,\end{aligned}$$

again using the equality of mixed partials.

Problems for Lecture 19

1. Find the curl of the following vector fields:

a) $\boldsymbol{F} = xy\boldsymbol{i} + yz\boldsymbol{j} + zx\boldsymbol{k}$;

b) $\boldsymbol{F} = yz\boldsymbol{i} + xz\boldsymbol{j} + xy\boldsymbol{k}$.

2. Consider a two-dimensional velocity field given by

$$\boldsymbol{u} = u_1(x,y)\boldsymbol{i} + u_2(x,y)\boldsymbol{j}.$$

Show that the vorticity $\boldsymbol{\omega} = \boldsymbol{\nabla} \times \boldsymbol{u}$ takes the form

$$\boldsymbol{\omega} = \omega_3(x,y)\boldsymbol{k}.$$

Determine ω_3 in terms of u_1 and u_2.

Lecture 20 Laplacian

The Laplacian differential operator, denoted as $\boldsymbol{\nabla} \cdot \boldsymbol{\nabla} = \nabla^2$ and pronounced as "del-squared", is given in Cartesian coordinates as

$$\nabla^2 = \frac{\partial^2}{\partial x^2} + \frac{\partial^2}{\partial y^2} + \frac{\partial^2}{\partial z^2}.$$

The Laplacian can be applied to either a scalar field or a vector field and results in a scalar field or a vector field, respectively. The Laplacian applied to a scalar field, $f = f(x, y, z)$, can be written as the divergence of the gradient, that is,

$$\nabla^2 f = \boldsymbol{\nabla} \cdot (\boldsymbol{\nabla} f) = \frac{\partial^2 f}{\partial x^2} + \frac{\partial^2 f}{\partial y^2} + \frac{\partial^2 f}{\partial z^2}.$$

The Laplacian applied to a vector field in Cartesian coordinates, acts on each component of the vector field separately. With $\boldsymbol{u} = u_1(x, y, z)\boldsymbol{i} + u_2(x, y, z)\boldsymbol{j} + u_3(x, y, z)\boldsymbol{k}$, we have

$$\nabla^2 \boldsymbol{u} = \nabla^2 u_1 \boldsymbol{i} + \nabla^2 u_2 \boldsymbol{j} + \nabla^2 u_3 \boldsymbol{k}.$$

The Laplacian appears in some classic partial differential equations. The Laplace equation, wave equation (with c the wave velocity), and diffusion equation (with D the diffusivity) all contain the Laplacian and are given, respectively, by

$$\nabla^2 \Phi = 0, \qquad \frac{\partial^2 \Phi}{\partial t^2} = c^2 \nabla^2 \Phi, \qquad \frac{\partial \Phi}{\partial t} = D \nabla^2 \Phi.$$

Example: Find the Laplacian of $f(x, y, z) = x^2 + y^2 + z^2$.

We have $\nabla^2 f = 2 + 2 + 2 = 6$.

Problems for Lecture 20

1. Compute $\nabla^2 \left(\frac{1}{r}\right)$ for $r \neq 0$. Here, $r = \sqrt{x^2 + y^2 + z^2}$.

Practice Quiz | The del operator

Let $\boldsymbol{r} = x\boldsymbol{i} + y\boldsymbol{j} + z\boldsymbol{k}$ and $r = \sqrt{x^2 + y^2 + z^2}$.

1. What is the value of $\boldsymbol{\nabla}\left(\frac{1}{r^2}\right)$?

a) $-\frac{\boldsymbol{r}}{r^3}$

b) $-\frac{2\boldsymbol{r}}{r^3}$

c) $-\frac{\boldsymbol{r}}{r^4}$

d) $-\frac{2\boldsymbol{r}}{r^4}$

2. Let $\boldsymbol{F} = \frac{\boldsymbol{r}}{r}$. The divergence $\boldsymbol{\nabla} \cdot \boldsymbol{F}$ is equal to

a) $\frac{1}{r}$

b) $\frac{2}{r}$

c) $\frac{1}{r^2}$

d) $\frac{2}{r^2}$

3. The curl of the position vector, $\boldsymbol{\nabla} \times \boldsymbol{r}$, is equal to

a) $\frac{\boldsymbol{r}}{r}$

b) 0

c) $\frac{(\boldsymbol{\nabla} \cdot \boldsymbol{r})\boldsymbol{r}}{r}$

d) $\boldsymbol{i} - \boldsymbol{j} + \boldsymbol{k}$

Lecture 21 Vector derivative identities

Let $f = f(\boldsymbol{r})$ be a scalar field and $\boldsymbol{u} = \boldsymbol{u}(\boldsymbol{r})$ and $\boldsymbol{v} = \boldsymbol{v}(\boldsymbol{r})$ be vector fields, where $\boldsymbol{u} = u_1\boldsymbol{i} + u_2\boldsymbol{j} + u_3\boldsymbol{k}$, and $\boldsymbol{v} = v_1\boldsymbol{i} + v_2\boldsymbol{j} + v_3\boldsymbol{k}$. Here, we will change notation and define the position vector to be $\boldsymbol{r} = x_1\boldsymbol{i} + x_2\boldsymbol{j} + x_3\boldsymbol{k}$ (instead of using the coordinates x, y and z). We have already shown that the curl of a gradient is zero, and the divergence of a curl is zero, that is,

$$\boldsymbol{\nabla} \times \boldsymbol{\nabla} f = 0, \qquad \boldsymbol{\nabla} \cdot (\boldsymbol{\nabla} \times \boldsymbol{u}) = 0.$$

Other sometimes useful vector derivative identities include

$$\begin{aligned}
\boldsymbol{\nabla} \times (\boldsymbol{\nabla} \times \boldsymbol{u}) &= \boldsymbol{\nabla}(\boldsymbol{\nabla} \cdot \boldsymbol{u}) - \nabla^2 \boldsymbol{u}, \\
\boldsymbol{\nabla} \cdot (f\boldsymbol{u}) &= \boldsymbol{u} \cdot \boldsymbol{\nabla} f + f \boldsymbol{\nabla} \cdot \boldsymbol{u}, \\
\boldsymbol{\nabla} \times (f\boldsymbol{u}) &= \boldsymbol{\nabla} f \times \boldsymbol{u} + f \boldsymbol{\nabla} \times \boldsymbol{u}, \\
\boldsymbol{\nabla}(\boldsymbol{u} \cdot \boldsymbol{v}) &= (\boldsymbol{u} \cdot \boldsymbol{\nabla})\boldsymbol{v} + (\boldsymbol{v} \cdot \boldsymbol{\nabla})\boldsymbol{u} + \boldsymbol{u} \times (\boldsymbol{\nabla} \times \boldsymbol{v}) + \boldsymbol{v} \times (\boldsymbol{\nabla} \times \boldsymbol{u}), \\
\boldsymbol{\nabla} \cdot (\boldsymbol{u} \times \boldsymbol{v}) &= \boldsymbol{v} \cdot (\boldsymbol{\nabla} \times \boldsymbol{u}) - \boldsymbol{u} \cdot (\boldsymbol{\nabla} \times \boldsymbol{v}), \\
\boldsymbol{\nabla} \times (\boldsymbol{u} \times \boldsymbol{v}) &= \boldsymbol{u}(\boldsymbol{\nabla} \cdot \boldsymbol{v}) - \boldsymbol{v}(\boldsymbol{\nabla} \cdot \boldsymbol{u}) + (\boldsymbol{v} \cdot \boldsymbol{\nabla})\boldsymbol{u} - (\boldsymbol{u} \cdot \boldsymbol{\nabla})\boldsymbol{v}.
\end{aligned}$$

Two of the identities make use of the del operator in the expression

$$\boldsymbol{u} \cdot \boldsymbol{\nabla} = u_1 \frac{\partial}{\partial x_1} + u_2 \frac{\partial}{\partial x_2} + u_3 \frac{\partial}{\partial x_3},$$

which acts on a scalar field as

$$\boldsymbol{u} \cdot \boldsymbol{\nabla} f = u_1 \frac{\partial f}{\partial x_1} + u_2 \frac{\partial f}{\partial x_2} + u_3 \frac{\partial f}{\partial x_3},$$

and acts on a vector field as

$$(\boldsymbol{u} \cdot \boldsymbol{\nabla})\boldsymbol{v} = (\boldsymbol{u} \cdot \boldsymbol{\nabla} v_1)\,\boldsymbol{i} + (\boldsymbol{u} \cdot \boldsymbol{\nabla} v_2)\,\boldsymbol{j} + (\boldsymbol{u} \cdot \boldsymbol{\nabla} v_3)\,\boldsymbol{k}.$$

In some of these identities, the parentheses are optional when the expression has only one possible interpretation. For example, it is common to see $(\boldsymbol{u} \cdot \boldsymbol{\nabla})\boldsymbol{v}$ written as $\boldsymbol{u} \cdot \boldsymbol{\nabla}\boldsymbol{v}$. The parentheses are mandatory when the expression can be interpreted in more than one way, for example $\boldsymbol{\nabla} \times \boldsymbol{u} \times \boldsymbol{v}$ could mean either $\boldsymbol{\nabla} \times (\boldsymbol{u} \times \boldsymbol{v})$ or $(\boldsymbol{\nabla} \times \boldsymbol{u}) \times \boldsymbol{v}$, and these two expressions are usually not equal.

Proof of all of these identities is most readily done by manipulating the Kronecker delta and Levi-Civita symbols, and I give an example in the next lecture.

Lecture 22 Vector derivative identities (proof)

To prove the vector derivative identities, we use component notation, the Einstein summation convention, the Levi-Civita symbol and the Kronecker delta. The ith component of the curl of a vector field is written using the Levi-Civita symbol as

$$(\boldsymbol{\nabla} \times \boldsymbol{u})_i = \epsilon_{ijk} \frac{\partial u_k}{\partial x_j};$$

and the divergence of a vector field is written as

$$\boldsymbol{\nabla} \cdot \boldsymbol{u} = \frac{\partial u_i}{\partial x_i}.$$

We will continue to make use of the usual Kronecker delta and Levi-Civita symbol identities.

As one example, I prove here the vector derivative identity

$$\boldsymbol{\nabla} \cdot (\boldsymbol{u} \times \boldsymbol{v}) = \boldsymbol{v} \cdot (\boldsymbol{\nabla} \times \boldsymbol{u}) - \boldsymbol{u} \cdot (\boldsymbol{\nabla} \times \boldsymbol{v}).$$

We have

$$\begin{aligned}
\boldsymbol{\nabla} \cdot (\boldsymbol{u} \times \boldsymbol{v}) &= \frac{\partial}{\partial x_i} \left(\epsilon_{ijk} u_j v_k \right) && \text{(write using component notation)} \\
&= \epsilon_{ijk} \frac{\partial u_j}{\partial x_i} v_k + \epsilon_{ijk} u_j \frac{\partial v_k}{\partial x_i} && \text{(product rule for the derivative)} \\
&= v_k \epsilon_{kij} \frac{\partial u_j}{\partial x_i} - u_j \epsilon_{jik} \frac{\partial v_k}{\partial x_i} && (\epsilon_{ijk} = \epsilon_{kij}, \quad \epsilon_{ijk} = -\epsilon_{jik}) \\
&= \boldsymbol{v} \cdot (\boldsymbol{\nabla} \times \boldsymbol{u}) - \boldsymbol{u} \cdot (\boldsymbol{\nabla} \times \boldsymbol{v}). && \text{(back to vector notation)}
\end{aligned}$$

The crucial step in the proof is the use of the product rule for the derivative. The rest of the proof just requires facility with the notation and the manipulation of the indices of the Levi-Civita symbol.

Problems for Lecture 22

1. Use the Kronecker delta, the Levi-Civita symbol and the Einstein summation convention, and the identities

$$\boldsymbol{a} \cdot \boldsymbol{b} = \delta_{ij} a_i b_j, \quad (\boldsymbol{a} \times \boldsymbol{b})_i = \epsilon_{ijk} a_j b_k, \quad \epsilon_{ijk}\epsilon_{ilm} = \delta_{jl}\delta_{km} - \delta_{jm}\delta_{kl},$$

to prove the following identities:

a) $\boldsymbol{\nabla} \cdot (f\boldsymbol{u}) = \boldsymbol{u} \cdot \boldsymbol{\nabla} f + f \boldsymbol{\nabla} \cdot \boldsymbol{u}$;

b) $\boldsymbol{\nabla} \times (\boldsymbol{\nabla} \times \boldsymbol{u}) = \boldsymbol{\nabla}(\boldsymbol{\nabla} \cdot \boldsymbol{u}) - \nabla^2 \boldsymbol{u}$.

2. Consider the vector differential equation for the position $\boldsymbol{r}$ of a fluid element subject to a velocity field $\boldsymbol{u}$ that depends on time t and position $\boldsymbol{r}$,

$$\frac{d\boldsymbol{r}}{dt} = \boldsymbol{u}(t, \boldsymbol{r}(t)),$$

where

$$\boldsymbol{r} = x_1\boldsymbol{i} + x_2\boldsymbol{j} + x_3\boldsymbol{k}, \qquad \boldsymbol{u} = u_1\boldsymbol{i} + u_2\boldsymbol{j} + u_3\boldsymbol{k}.$$

a) Write down the differential equations for dx_1/dt, dx_2/dt and dx_3/dt;

b) Use the chain rule to determine formulas for d^2x_1/dt^2, d^2x_2/dt^2 and d^2x_3/dt^2;

c) Write your solution for $d^2\boldsymbol{r}/dt^2$ as a vector equation using the $\boldsymbol{\nabla}$ differential operator.

Lecture 23 Electromagnetic waves

Maxwell's equations for the electric field $\boldsymbol{E}$ and the magnetic field $\boldsymbol{B}$ in the vacuum of free space are most simply written using the del operator, and are given by

$$\boldsymbol{\nabla} \cdot \boldsymbol{E} = 0, \qquad \boldsymbol{\nabla} \cdot \boldsymbol{B} = 0, \qquad \boldsymbol{\nabla} \times \boldsymbol{E} = -\frac{\partial \boldsymbol{B}}{\partial t}, \qquad \boldsymbol{\nabla} \times \boldsymbol{B} = \mu_0 \epsilon_0 \frac{\partial \boldsymbol{E}}{\partial t}.$$

Here I use the SI units familiar to engineering students, where the constants ϵ_0 and μ_0 are called the permittivity and permeability of free space, respectively.

From the four Maxwell's equations, we would like to obtain a single equation for $\boldsymbol{E}$. To do so, we can make use of the curl of the curl identity

$$\boldsymbol{\nabla} \times (\boldsymbol{\nabla} \times \boldsymbol{E}) = \boldsymbol{\nabla}(\boldsymbol{\nabla} \cdot \boldsymbol{E}) - \nabla^2 \boldsymbol{E}.$$

To obtain an equation for $\boldsymbol{E}$, we take the curl of the third Maxwell's equation and commute the time and space derivatives

$$\boldsymbol{\nabla} \times (\boldsymbol{\nabla} \times \boldsymbol{E}) = -\frac{\partial}{\partial t}(\boldsymbol{\nabla} \times \boldsymbol{B}).$$

We apply the curl of the curl identity to obtain

$$\boldsymbol{\nabla}(\boldsymbol{\nabla} \cdot \boldsymbol{E}) - \nabla^2 \boldsymbol{E} = -\frac{\partial}{\partial t}(\boldsymbol{\nabla} \times \boldsymbol{B}),$$

and then apply the first Maxwell's equation to the left-hand-side, and the fourth Maxwell's equation to the right-hand-side. Rearranging terms, we obtain the three-dimensional wave equation given by

$$\frac{\partial^2 \boldsymbol{E}}{\partial t^2} = c^2 \nabla^2 \boldsymbol{E},$$

with c the wave speed given by $c = 1/\sqrt{\mu_0 \epsilon_0} \approx 3 \times 10^8$ m/s. This is, of course, the speed of light in vacuum.

Problems for Lecture 23

1. Derive the wave equation for the magnetic field B.

Practice Quiz | Vector calculus algebra

1. Let $\boldsymbol{u}$ and $\boldsymbol{v}$ be vector fields. Using the vector derivative identity $\boldsymbol{\nabla}(\boldsymbol{u} \cdot \boldsymbol{v}) = (\boldsymbol{u} \cdot \boldsymbol{\nabla})\boldsymbol{v} + (\boldsymbol{v} \cdot \boldsymbol{\nabla})\boldsymbol{u} + \boldsymbol{u} \times (\boldsymbol{\nabla} \times \boldsymbol{v}) + \boldsymbol{v} \times (\boldsymbol{\nabla} \times \boldsymbol{u})$, which of the following identities is valid?

a) $\dfrac{1}{2}\boldsymbol{\nabla}(\boldsymbol{u} \cdot \boldsymbol{u}) = \boldsymbol{u} \times (\boldsymbol{\nabla} \times \boldsymbol{u}) + (\boldsymbol{u} \cdot \boldsymbol{\nabla})\boldsymbol{u}$

b) $\dfrac{1}{2}\boldsymbol{\nabla}(\boldsymbol{u} \cdot \boldsymbol{u}) = \boldsymbol{u} \times (\boldsymbol{\nabla} \times \boldsymbol{u}) - (\boldsymbol{u} \cdot \boldsymbol{\nabla})\boldsymbol{u}$

c) $\dfrac{1}{2}\boldsymbol{\nabla}(\boldsymbol{u} \cdot \boldsymbol{u}) = (\boldsymbol{\nabla} \times \boldsymbol{u}) \times \boldsymbol{u} + (\boldsymbol{u} \cdot \boldsymbol{\nabla})\boldsymbol{u}$

d) $\dfrac{1}{2}\boldsymbol{\nabla}(\boldsymbol{u} \cdot \boldsymbol{u}) = (\boldsymbol{\nabla} \times \boldsymbol{u}) \times \boldsymbol{u} - (\boldsymbol{u} \cdot \boldsymbol{\nabla})\boldsymbol{u}$

2. Let $\boldsymbol{u}$ be a vector field and f be a scalar field. Which of the following expressions is not always zero?

a) $\boldsymbol{\nabla} \times (\boldsymbol{\nabla} f)$

b) $\boldsymbol{\nabla} \cdot (\boldsymbol{\nabla} \times \boldsymbol{u})$

c) $\boldsymbol{\nabla} \cdot (\boldsymbol{\nabla} f)$

d) $\boldsymbol{\nabla} \times (\boldsymbol{\nabla}(\boldsymbol{\nabla} \cdot \boldsymbol{u}))$

3. Suppose the electric field is given by $\boldsymbol{E}(\boldsymbol{r},t) = \sin(z - ct)\boldsymbol{i}$. Then which of the following is a valid free-space solution for the magnetic field $\boldsymbol{B} = \boldsymbol{B}(\boldsymbol{r},t)$?

a) $\boldsymbol{B}(\boldsymbol{r},t) = \dfrac{1}{c}\sin(z - ct)\boldsymbol{i}$

b) $\boldsymbol{B}(\boldsymbol{r},t) = \dfrac{1}{c}\sin(z - ct)\boldsymbol{j}$

c) $\boldsymbol{B}(\boldsymbol{r},t) = \dfrac{1}{c}\sin(x - ct)\boldsymbol{i}$

d) $\boldsymbol{B}(\boldsymbol{r},t) = \dfrac{1}{c}\sin(x - ct)\boldsymbol{j}$

Week III

Integration and Curvilinear Coordinates

In this week's lectures, we learn about multidimensional integration. The important technique of using curvilinear coordinates, namely polar coordinates in two dimensions, and cylindrical and spherical coordinates in three dimensions, is used to simplify problems with circular, cylindrical or spherical symmetry. Differential operators in curvilinear coordinates are derived. The change of variables formula for multidimensional integrals using the Jacobian of the transformation is explained.

Lecture 24 Double and triple integrals

Double and triple integrals, written as

$$\int_A f dA = \iint_A f(x,y)\,dx\,dy, \qquad \int_V f dV = \iiint_V f(x,y,z)\,dx\,dy\,dz,$$

are the limits of the sums of $\Delta x \Delta y$ (or $\Delta x \Delta y \Delta z$) multiplied by the integrand. A single integral is the area under a curve $y = f(x)$; a double integral is the volume under a surface $z = f(x,y)$. A triple integral is used, for example, to find the mass of an object by integrating over its density.

To perform a double or triple integral, the correct limits of the integral need to be determined, and the integral is performed as two (or three) single integrals. For example, an integration over a rectangle in the x-y plane can be written as either

$$\int_{y_0}^{y_1} \int_{x_0}^{x_1} f(x,y)\,dx\,dy \quad \text{or} \quad \int_{x_0}^{x_1} \int_{y_0}^{y_1} f(x,y)\,dy\,dx.$$

In the first double integral, the x integration is done first (holding y fixed), and the y integral is done second. In the second double integral, the order of integration is reversed. Either order of integration will give the same result.

Example: Compute the volume of the surface $z = x^2 y$ above the x-y plane with base given by a unit square with vertices $(0,0)$, $(1,0)$, $(1,1)$, and $(0,1)$.

To find the volume, we integrate $z = x^2 y$ over its base. The integral over the unit square is given by either of the double integrals

$$\int_0^1 \int_0^1 x^2 y\,dx\,dy \quad \text{or} \quad \int_0^1 \int_0^1 x^2 y\,dy\,dx.$$

The respective calculations are

$$\int_0^1 \int_0^1 x^2 y\,dx\,dy = \int_0^1 \left(\frac{x^3 y}{3} \bigg|_{x=0}^{x=1} \right) dy = \frac{1}{3} \int_0^1 y\,dy = \frac{1}{3}\frac{y^2}{2} \bigg|_{y=0}^{y=1} = \frac{1}{6};$$

$$\int_0^1 \int_0^1 x^2 y\,dy\,dx = \int_0^1 \left(\frac{x^2 y^2}{2} \bigg|_{y=0}^{y=1} \right) dx = \frac{1}{2} \int_0^1 x^2\,dy = \frac{1}{2}\frac{x^3}{3} \bigg|_{x=0}^{x=1} = \frac{1}{6}.$$

In this case, an even simpler integration method separates the x and y dependence and writes

$$\int_0^1 \int_0^1 x^2 y\,dx\,dy = \int_0^1 x^2\,dx \int_0^1 y\,dy = \left(\frac{1}{3}\right)\left(\frac{1}{2}\right) = \frac{1}{6}.$$

Problems for Lecture 24

1. A cube has edge length L, with mass density increasing linearly from ρ_1 to ρ_2 from one face of the cube to the opposite face. By solving a triple integral, compute the mass of the cube in terms of L, ρ_1 and ρ_2.

Lecture 25 Example: Double integral with triangle base

Example: Compute the volume of the surface $z = x^2y$ above the x-y plane with base given by a right triangle with vertices $(0,0)$, $(1,0)$, $(0,1)$.

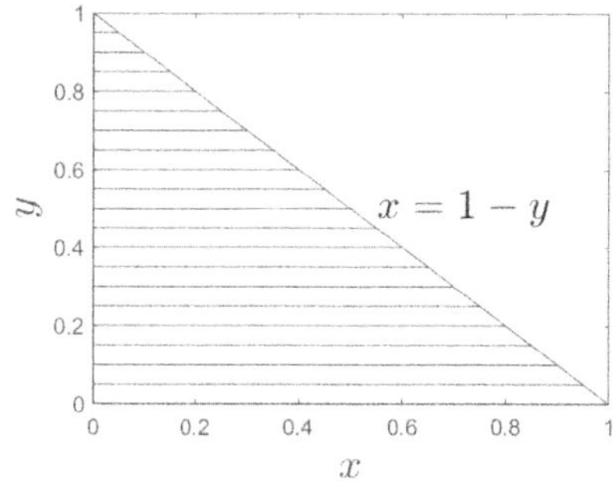

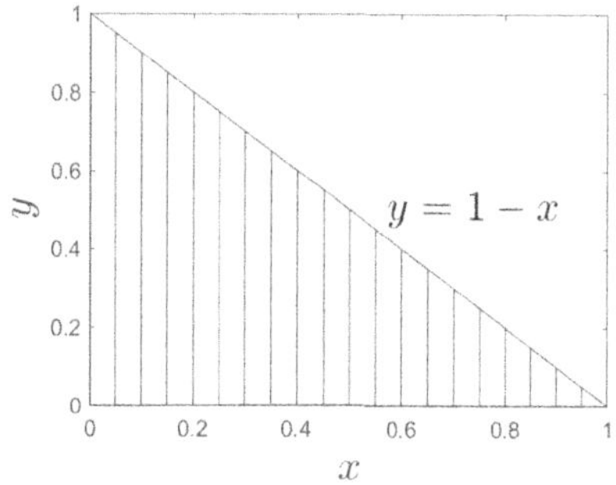

The left figure illustrates the integral over x first and y second; and the right figure illustrates the integral over y first and x second. These are the two double integrals,

$$\int_0^1 \int_0^{1-y} x^2y\,dx\,dy, \quad \text{or} \quad \int_0^1 \int_0^{1-x} x^2y\,dy\,dx.$$

The respective calculations are

$$\int_0^1 \int_0^{1-y} x^2y\,dx\,dy = \int_0^1 \left(\frac{x^3y}{3}\bigg|_{x=0}^{x=1-y} \right) dy = \frac{1}{3}\int_0^1 (1-y)^3 y\,dy$$

$$= \frac{1}{3}\int_0^1 (y - 3y^2 + 3y^3 - y^4)\,dy = \frac{1}{3}\left(\frac{y^2}{2} - y^3 + \frac{3y^4}{4} - \frac{y^5}{5} \right)\bigg|_0^1$$

$$= \frac{1}{3}\left(\frac{1}{2} - 1 + \frac{3}{4} - \frac{1}{5} \right) = \frac{1}{60};$$

$$\int_0^1 \int_0^{1-y} x^2y\,dy\,dx = \int_0^1 \left(\frac{x^2y^2}{2}\bigg|_{y=0}^{y=1-x} \right) dx = \frac{1}{2}\int_0^1 x^2(1-x)^2\,dx$$

$$= \frac{1}{2}\int_0^1 (x^2 - 2x^3 + x^4)\,dx = \frac{1}{2}\left(\frac{x^3}{3} - \frac{x^4}{2} + \frac{x^5}{5} \right)\bigg|_0^1 = \frac{1}{2}\left(\frac{1}{3} - \frac{1}{2} + \frac{1}{5} \right) = \frac{1}{60}.$$

Problems for Lecture 25

1. Compute the volume of the surface $z = x^2y$ above the x-y plane with base given by a parallelogram with vertices $(0,0)$, $(1,0)$, $(4/3,1)$ and $(1/3,1)$.

Practice Quiz | Multi-dimensional integration

1. The volume of the surface $z = xy$ above the x-y plane with base given by a unit square with vertices $(0,0)$, $(1,0)$, $(1,1)$, and $(0,1)$ is equal to

a) $\frac{1}{5}$

b) $\frac{1}{4}$

c) $\frac{1}{3}$

d) $\frac{1}{2}$

2. A cube has edge length of 1 cm, with mass density increasing linearly from $1\,\text{g/cm}^3$ to $2\,\text{g/cm}^3$ from one face of the cube to the opposite face. The mass of the cube is given by

a) 3.0 g

b) 1.5 g

c) 1.33 g

d) 1.0 g

3. The volume of the surface $z = xy$ above the x-y plane with base given by the triangle with vertices $(0,0)$, $(1,1)$, and $(2,0)$ is equal to

a) $\frac{1}{6}$

b) $\frac{1}{5}$

c) $\frac{1}{4}$

d) $\frac{1}{3}$

Lecture 26 | Polar coordinates (gradient)

In two dimensions, polar coordinates are the most commonly used curvilinear coordinate system. The relationship between Cartesian and polar coordinates is given by

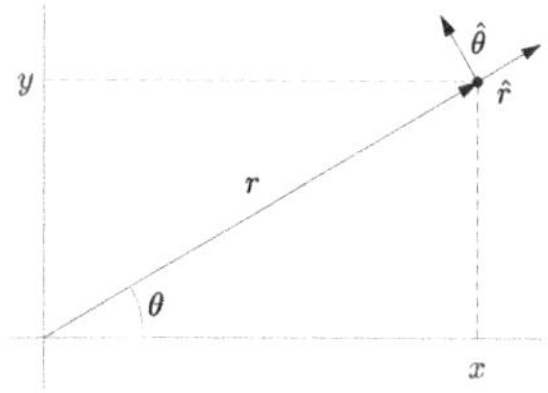

$$x = r\cos\theta, \qquad y = r\sin\theta.$$

The unit vectors $\hat{\boldsymbol{r}}$ and $\hat{\boldsymbol{\theta}}$ are defined to be orthogonal and in the direction of increasing r and θ, respectively, and the position vector is given by $\boldsymbol{r} = r\hat{\boldsymbol{r}}$. The $\hat{\boldsymbol{r}}$-$\hat{\boldsymbol{\theta}}$ unit vectors are rotated an angle θ from the $\boldsymbol{i}$-$\boldsymbol{j}$ unit vectors. Trigonometry shows that

$$\hat{\boldsymbol{r}} = \cos\theta\boldsymbol{i} + \sin\theta\boldsymbol{j}, \qquad \hat{\boldsymbol{\theta}} = -\sin\theta\boldsymbol{i} + \cos\theta\boldsymbol{j}.$$

Notice that the direction of the unit vectors in polar coordinates is not fixed, but depends on their position. Here, $\hat{\boldsymbol{r}} = \hat{\boldsymbol{r}}(\theta)$ and $\hat{\boldsymbol{\theta}} = \hat{\boldsymbol{\theta}}(\theta)$, and their derivatives are given by

$$\frac{d\hat{\boldsymbol{r}}}{d\theta} = \hat{\boldsymbol{\theta}}, \qquad \frac{d\hat{\boldsymbol{\theta}}}{d\theta} = -\hat{\boldsymbol{r}}.$$

Now, consider a two-dimensional scalar field given by $f = f(r,\theta)$. By using the definition of the two-dimensional gradient or by applying the chain rule, the differential df can be written as

$$df = \boldsymbol{\nabla} f \cdot d\boldsymbol{r} \qquad \text{or} \qquad df = \frac{\partial f}{\partial r}dr + \frac{\partial f}{\partial \theta}d\theta.$$

We have

$$d\boldsymbol{r} = d(r\hat{\boldsymbol{r}}) = \hat{\boldsymbol{r}}dr + r\frac{d\hat{\boldsymbol{r}}}{d\theta}d\theta = \hat{\boldsymbol{r}}dr + r\hat{\boldsymbol{\theta}}d\theta,$$

and equating the two forms for df results in

$$\boldsymbol{\nabla} f \cdot (\hat{\boldsymbol{r}}dr + r\hat{\boldsymbol{\theta}}d\theta) = \frac{\partial f}{\partial r}dr + \frac{\partial f}{\partial \theta}d\theta.$$

By inspection, we see that

$$\boldsymbol{\nabla} f = \frac{\partial f}{\partial r}\hat{\boldsymbol{r}} + \frac{1}{r}\frac{\partial f}{\partial \theta}\hat{\boldsymbol{\theta}},$$

so that the gradient operator, in polar coordinates, is given by

$$\boldsymbol{\nabla} = \hat{\boldsymbol{r}}\frac{\partial}{\partial r} + \hat{\boldsymbol{\theta}}\frac{1}{r}\frac{\partial}{\partial \theta}.$$

Problems for Lecture 26

1. The inverse of a two-by-two matrix is given by

$$\begin{pmatrix} a & b \\ c & d \end{pmatrix}^{-1} = \frac{1}{ad - bc} \begin{pmatrix} d & -b \\ -c & a \end{pmatrix}.$$

Given

$$\hat{\boldsymbol{r}} = \cos\theta \boldsymbol{i} + \sin\theta \boldsymbol{j}, \qquad \hat{\boldsymbol{\theta}} = -\sin\theta \boldsymbol{i} + \cos\theta \boldsymbol{j},$$

invert a two-by-two matrix to solve for $\boldsymbol{i}$ and $\boldsymbol{j}$.

2. Cartesian coordinates are related to polar coordinates by the equations $x = r\cos\theta$ and $y = r\sin\theta$.

a) Let $f = f(x(r,\theta), y(r,\theta))$. Using the chain rule, show that

$$\frac{\partial f}{\partial r} = \cos\theta \frac{\partial f}{\partial x} + \sin\theta \frac{\partial f}{\partial y}, \qquad \frac{\partial f}{\partial \theta} = -r\sin\theta \frac{\partial f}{\partial x} + r\cos\theta \frac{\partial f}{\partial y}.$$

b) Invert the result of Part (a) to show that

$$\frac{\partial}{\partial x} = \cos\theta \frac{\partial}{\partial r} - \frac{\sin\theta}{r} \frac{\partial}{\partial \theta}, \qquad \frac{\partial}{\partial y} = \sin\theta \frac{\partial}{\partial r} + \frac{\cos\theta}{r} \frac{\partial}{\partial \theta}.$$

3. Determine expressions for $r\hat{\boldsymbol{r}}$ and $r\hat{\boldsymbol{\theta}}$ in Cartesian coordinates.

Lecture 27 Polar coordinates (divergence & curl)

Define a vector field in polar coordinates to be

$$\boldsymbol{u} = u_r(r,\theta)\hat{\boldsymbol{r}} + u_\theta(r,\theta)\hat{\boldsymbol{\theta}}.$$

To compute the divergence and curl in polar coordinates, we will use $\hat{\boldsymbol{r}} = \hat{\boldsymbol{r}}(\theta)$ and $\hat{\boldsymbol{\theta}} = \hat{\boldsymbol{\theta}}(\theta)$, and

$$\frac{d\hat{\boldsymbol{r}}}{d\theta} = \hat{\boldsymbol{\theta}}, \qquad \frac{d\hat{\boldsymbol{\theta}}}{d\theta} = -\hat{\boldsymbol{r}}, \qquad \boldsymbol{\nabla} = \hat{\boldsymbol{r}}\frac{\partial}{\partial r} + \hat{\boldsymbol{\theta}}\frac{1}{r}\frac{\partial}{\partial \theta}.$$

The polar unit vectors also satisfy

$$\hat{\boldsymbol{r}}\cdot\hat{\boldsymbol{r}} = \hat{\boldsymbol{\theta}}\cdot\hat{\boldsymbol{\theta}} = 1, \qquad \hat{\boldsymbol{r}}\cdot\hat{\boldsymbol{\theta}} = \hat{\boldsymbol{\theta}}\cdot\hat{\boldsymbol{r}} = 0,$$

$$\hat{\boldsymbol{r}}\times\hat{\boldsymbol{r}} = \hat{\boldsymbol{\theta}}\times\hat{\boldsymbol{\theta}} = 0, \quad \hat{\boldsymbol{r}}\times\hat{\boldsymbol{\theta}} = -(\hat{\boldsymbol{\theta}}\times\hat{\boldsymbol{r}}) = \hat{\boldsymbol{k}},$$

where $\hat{\boldsymbol{k}}$ is the standard Cartesian unit vector pointing along the z-axis.

The divergence is computed from

$$\begin{aligned}
\boldsymbol{\nabla}\cdot\boldsymbol{u} &= \left(\hat{\boldsymbol{r}}\frac{\partial}{\partial r} + \hat{\boldsymbol{\theta}}\frac{1}{r}\frac{\partial}{\partial \theta}\right)\cdot\left(u_r\hat{\boldsymbol{r}} + u_\theta\hat{\boldsymbol{\theta}}\right) \\
&= \hat{\boldsymbol{r}}\cdot\frac{\partial}{\partial r}\left(u_r\hat{\boldsymbol{r}}\right) + \hat{\boldsymbol{r}}\cdot\frac{\partial}{\partial r}\left(u_\theta\hat{\boldsymbol{\theta}}\right) + \hat{\boldsymbol{\theta}}\cdot\frac{1}{r}\frac{\partial}{\partial \theta}\left(u_r\hat{\boldsymbol{r}}\right) + \hat{\boldsymbol{\theta}}\cdot\frac{1}{r}\frac{\partial}{\partial \theta}\left(u_\theta\hat{\boldsymbol{\theta}}\right) \\
&= \frac{\partial u_r}{\partial r} + 0 + \frac{1}{r}u_r + \frac{1}{r}\frac{\partial u_\theta}{\partial \theta} \\
&= \frac{1}{r}\frac{\partial}{\partial r}\left(ru_r\right) + \frac{1}{r}\frac{\partial u_\theta}{\partial \theta}.
\end{aligned}$$

And the curl is computed from

$$\begin{aligned}
\boldsymbol{\nabla}\times\boldsymbol{u} &= \left(\hat{\boldsymbol{r}}\frac{\partial}{\partial r} + \hat{\boldsymbol{\theta}}\frac{1}{r}\frac{\partial}{\partial \theta}\right)\times\left(u_r\hat{\boldsymbol{r}} + u_\theta\hat{\boldsymbol{\theta}}\right) \\
&= \hat{\boldsymbol{r}}\times\frac{\partial}{\partial r}\left(u_r\hat{\boldsymbol{r}}\right) + \hat{\boldsymbol{r}}\times\frac{\partial}{\partial r}\left(u_\theta\hat{\boldsymbol{\theta}}\right) + \hat{\boldsymbol{\theta}}\times\frac{1}{r}\frac{\partial}{\partial \theta}\left(u_r\hat{\boldsymbol{r}}\right) + \hat{\boldsymbol{\theta}}\times\frac{1}{r}\frac{\partial}{\partial \theta}\left(u_\theta\hat{\boldsymbol{\theta}}\right) \\
&= 0 + \frac{\partial u_\theta}{\partial r}\left(\hat{\boldsymbol{r}}\times\hat{\boldsymbol{\theta}}\right) + \frac{1}{r}\frac{\partial u_r}{\partial \theta}\left(\hat{\boldsymbol{\theta}}\times\hat{\boldsymbol{r}}\right) - \frac{1}{r}u_\theta\left(\hat{\boldsymbol{\theta}}\times\hat{\boldsymbol{r}}\right) \\
&= \hat{\boldsymbol{k}}\left(\frac{1}{r}\frac{\partial}{\partial r}\left(ru_\theta\right) - \frac{1}{r}\frac{\partial u_r}{\partial \theta}\right).
\end{aligned}$$

Note that the curl of a two-dimensional vector field defined in a plane points in the direction perpendicular to the plane following the right-hand rule.

Problems for Lecture 27

1. Let $\boldsymbol{u}$ be a two-dimensional vector field given in polar coordinates by

$$\boldsymbol{u} = \frac{1}{r}\left(k_1\hat{\boldsymbol{r}} + k_2\hat{\boldsymbol{\theta}}\right),$$

where k_1 and k_2 are constants. For $r \neq 0$, determine $\boldsymbol{\nabla} \cdot \boldsymbol{u}$ and $\boldsymbol{\nabla} \times \boldsymbol{u}$.

Lecture 28 Polar coordinates (Laplacian)

The two-dimensional Laplacian operator in polar coordinates can be computed using

$$\nabla^2 = (\boldsymbol{\nabla} \cdot \boldsymbol{\nabla}), \qquad \text{where} \qquad \boldsymbol{\nabla} = \hat{\boldsymbol{r}}\frac{\partial}{\partial r} + \hat{\boldsymbol{\theta}}\frac{1}{r}\frac{\partial}{\partial \theta}.$$

We have

$$\begin{aligned}
\nabla^2 &= \left(\hat{\boldsymbol{r}}\frac{\partial}{\partial r} + \hat{\boldsymbol{\theta}}\frac{1}{r}\frac{\partial}{\partial \theta}\right) \cdot \left(\hat{\boldsymbol{r}}\frac{\partial}{\partial r} + \hat{\boldsymbol{\theta}}\frac{1}{r}\frac{\partial}{\partial \theta}\right) \\
&= \hat{\boldsymbol{r}} \cdot \frac{\partial}{\partial r}\left(\hat{\boldsymbol{r}}\frac{\partial}{\partial r}\right) + \hat{\boldsymbol{r}} \cdot \frac{\partial}{\partial r}\left(\hat{\boldsymbol{\theta}}\frac{1}{r}\frac{\partial}{\partial \theta}\right) + \hat{\boldsymbol{\theta}} \cdot \frac{1}{r}\frac{\partial}{\partial \theta}\left(\hat{\boldsymbol{r}}\frac{\partial}{\partial r}\right) + \hat{\boldsymbol{\theta}} \cdot \frac{1}{r}\frac{\partial}{\partial \theta}\left(\hat{\boldsymbol{\theta}}\frac{1}{r}\frac{\partial}{\partial \theta}\right) \\
&= \frac{\partial^2}{\partial r^2} + 0 + \frac{1}{r}\frac{\partial}{\partial r} + \frac{1}{r^2}\frac{\partial^2}{\partial \theta^2} \\
&= \frac{1}{r}\frac{\partial}{\partial r}\left(r\frac{\partial}{\partial r}\right) + \frac{1}{r^2}\frac{\partial^2}{\partial \theta^2}.
\end{aligned}$$

The Laplacian operator in polar coordinates can be applied to either a scalar or vector field. When applied to a vector field, one needs to differentiate the unit vectors with respect to θ.

Problems for Lecture 28

1. Consider the fluid flow through a pipe of circular cross-section radius R with a constant pressure gradient along the pipe length. Define the z-axis to be the symmetry axis down the center of the pipe in the direction of the flowing fluid. The velocity field in a steady flow then takes the form $\boldsymbol{u} = u(r)\boldsymbol{k}$, where $r = \sqrt{x^2 + y^2}$ and u satisfies the Navier-Stokes equation given by

$$\nabla^2 u = -\frac{G}{\nu\rho}.$$

Here, G is the pressure gradient, ν is the kinematic viscosity, and ρ is the fluid density, all assumed to be constant. You may further assume that the interior surface of the pipe has no slip so that the fluid velocity is zero when $r = R$. Solve for the velocity field $u(r)$ in the pipe's cross section using the polar coordinate form for the Laplacian. What is the maximum value of the velocity?

Lecture 29 | Example: Central force

A central force is a force acting on a point mass and pointing directly towards a fixed point in space. The origin of the coordinate system is chosen at this fixed point, and the axis orientated such that the initial position and velocity of the mass lies in the x-y plane. If the central force is the only relevant force, the subsequent motion of the mass is then two dimensional and polar coordinates can be employed.

The position vector of the point mass in polar coordinates is given by

$$\boldsymbol{r} = r\hat{\boldsymbol{r}}.$$

The velocity and acceleration of the point mass is obtained by differentiating $\boldsymbol{r}$ with respect to time. The algebra is made complicated because the unit vectors $\hat{\boldsymbol{r}} = \hat{\boldsymbol{r}}(\theta(t))$ and $\hat{\boldsymbol{\theta}} = \hat{\boldsymbol{\theta}}(\theta(t))$ are also functions of time. When differentiating, we will need to use the chain rule in the form

$$\frac{d\hat{\boldsymbol{r}}}{dt} = \frac{d\hat{\boldsymbol{r}}}{d\theta}\frac{d\theta}{dt} = \frac{d\theta}{dt}\hat{\boldsymbol{\theta}}, \qquad \frac{d\hat{\boldsymbol{\theta}}}{dt} = \frac{d\hat{\boldsymbol{\theta}}}{d\theta}\frac{d\theta}{dt} = -\frac{d\theta}{dt}\hat{\boldsymbol{r}}.$$

As is customary, we make use of the dot notation for the time derivative. For example, $\dot{x} = dx/dt$ and $\ddot{x} = d^2x/dt^2$.

The velocity of the point mass is then given by

$$\dot{\boldsymbol{r}} = \dot{r}\hat{\boldsymbol{r}} + r\frac{d\hat{\boldsymbol{r}}}{dt} = \dot{r}\hat{\boldsymbol{r}} + r\dot{\theta}\hat{\boldsymbol{\theta}};$$

and the acceleration is given by

$$\begin{aligned}\ddot{\boldsymbol{r}} &= \ddot{r}\hat{\boldsymbol{r}} + \dot{r}\frac{d\hat{\boldsymbol{r}}}{dt} + \dot{r}\dot{\theta}\hat{\boldsymbol{\theta}} + r\ddot{\theta}\hat{\boldsymbol{\theta}} + r\dot{\theta}\frac{d\hat{\boldsymbol{\theta}}}{dt} \\ &= (\ddot{r} - r\dot{\theta}^2)\hat{\boldsymbol{r}} + (2\dot{r}\dot{\theta} + r\ddot{\theta})\hat{\boldsymbol{\theta}}.\end{aligned}$$

A central force can be written as $\boldsymbol{F} = -f\hat{\boldsymbol{r}}$, where usually $f = f(r)$. Newton's equation, $m\ddot{\boldsymbol{r}} = \boldsymbol{F}$, then separates into the two component equations,

$$m(\ddot{r} - r\dot{\theta}^2) = -f, \qquad m(2\dot{r}\dot{\theta} + r\ddot{\theta}) = 0.$$

The second equation is usually expressed as conservation of angular momentum, and after multiplication by r, is written in the form

$$\frac{d}{dt}(mr^2\dot{\theta}) = 0, \quad \text{or} \quad mr^2\dot{\theta} = \text{constant}.$$

Problems for Lecture 29

1. The angular momentum $\boldsymbol{l}$ of a point mass m relative to an origin is defined as

$$\boldsymbol{l} = \boldsymbol{r} \times \boldsymbol{p},$$

where $\boldsymbol{r}$ is the position vector of the mass and $\boldsymbol{p} = m\dot{\boldsymbol{r}}$ is the momentum of the mass. Show that

$$|\boldsymbol{l}| = mr^2|\dot{\theta}|.$$

2. Prove that

$$m\boldsymbol{v} \cdot \frac{d\boldsymbol{v}}{dt} = \frac{d}{dt}\left(\frac{1}{2}m|\boldsymbol{v}|^2\right).$$

This result will be used to derive the conservation of energy.

Lecture 30 | Change of variables (single integral)

A double (or triple) integral written in Cartesian coordinates, can sometimes be more easily computed by changing coordinate systems. To do so, we need to derive a change of variables formula.

It is illuminating to first revisit the change-of-variables formula for single integrals. Consider the integral

$$I = \int_{x_0}^{x_f} f(x)\, dx.$$

Let $u(s)$ be a differentiable and invertible function. We can change variables in this integral by letting $x = u(s)$ so that $dx = u'(s)\, ds$. The integral in the new variable s then becomes

$$I = \int_{u^{-1}(x_0)}^{u^{-1}(x_f)} f(u(s))u'(s)\, ds.$$

The key piece for us here is the transformation of the infinitesimal length $dx = u'(s)\, ds$.

We can be more concrete by examining a specific transformation. Consider the calculation of the area of a circle of radius R, given by the integral

$$A = 4\int_0^R \sqrt{R^2 - x^2}\, dx.$$

To more easily perform this integral, we can let $x = R\cos\theta$ so that $dx = -R\sin\theta\, d\theta$. The integral then becomes

$$A = 4R^2 \int_0^{\pi/2} \sqrt{1-\cos^2\theta}\sin\theta\, d\theta = 4R^2 \int_0^{\pi/2} \sin^2\theta\, d\theta,$$

which can be done using the double-angle formula to yield $A = \pi R^2$.

Lecture 31 Change of variables (double integral)

We consider the double integral

$$I = \iint_A f(x,y)\,dx\,dy.$$

We would like to change variables from (x,y) to (s,t). For simplicity, we will write this change of variables as $x = x(s,t)$ and $y = y(s,t)$. The region A in the x-y domain transforms into a region A' in the s-t domain, and the integrand becomes a function of the new variables s and t by the substitution $f(x,y) = f(x(s,t), y(s,t))$. We now consider how to transform the infinitesimal area $dx\,dy$.

The transformation of $dx\,dy$ is obtained by considering how an infinitesimal rectangle is transformed into an infinitesimal parallelogram, and how the area of the two are related by the absolute value of a determinant. The main result, which we do not derive here, is given by

$$dx\,dy = |\det(J)|\,ds\,dt,$$

where J is called the Jacobian of the transformation, and is the matrix defined as

$$J = \frac{\partial(x,y)}{\partial(s,t)} = \begin{pmatrix} \partial x/\partial s & \partial x/\partial t \\ \partial y/\partial s & \partial y/\partial t \end{pmatrix}.$$

To be more concrete, we again calculate the area of a circle. Here, using a two-dimensional integral, the area of a circle can be written as

$$A = \iint_A dx\,dy,$$

where the integral subscript A denotes the region in the x-y plane that defines the circle. To perform this integral, we can change from Cartesian to polar coordinates. Let

$$x = r\cos\theta, \qquad y = r\sin\theta.$$

We have

$$dx\,dy = \left|\det\begin{pmatrix} \partial x/\partial r & \partial x/\partial\theta \\ \partial y/\partial r & \partial y/\partial\theta \end{pmatrix}\right| dr\,d\theta = \left|\det\begin{pmatrix} \cos\theta & -r\sin\theta \\ \sin\theta & r\cos\theta \end{pmatrix}\right| dr\,d\theta = r\,dr\,d\theta.$$

The region in the r-θ plane that defines the circle is $0 \leq r \leq R$ and $0 \leq \theta \leq 2\pi$. The integral then becomes

$$A = \int_0^{2\pi}\int_0^R r\,dr\,d\theta = \int_0^{2\pi} d\theta \int_0^R r\,dr = \pi R^2.$$

Problems for Lecture 31

1. The mass density of a flat object can be specified by $\sigma = \sigma(x, y)$, with units of mass per unit area. The total mass of the object is found from the double integral

$$M = \int\int_A \sigma(x, y)\, dx\, dy.$$

Suppose a circular disk of radius R has mass density ρ_0 at its center and ρ_1 at its edge, and its density is a linear function of the distance from the center. Find the total mass of the disk.

2. Compute the Gaussian integral given by $I = \int_{-\infty}^{\infty} e^{-x^2}\, dx$. Use the well-known trick

$$I^2 = \left(\int_{-\infty}^{\infty} e^{-x^2}\, dx\right)^2 = \int_{-\infty}^{\infty} e^{-x^2}\, dx \int_{-\infty}^{\infty} e^{-y^2}\, dy = \int_{-\infty}^{\infty}\int_{-\infty}^{\infty} e^{-(x^2+y^2)}\, dx\, dy.$$

Practice Quiz | Polar coordinates

1. $r\hat{\boldsymbol{\theta}}$ is equal to

a) $x\boldsymbol{i} + y\boldsymbol{j}$

b) $x\boldsymbol{i} - y\boldsymbol{j}$

c) $y\boldsymbol{i} + x\boldsymbol{j}$

d) $-y\boldsymbol{i} + x\boldsymbol{j}$

2. $\dfrac{d\hat{\boldsymbol{\theta}}}{d\theta}$ is equal to

a) $\hat{\boldsymbol{r}}$

b) $-\hat{\boldsymbol{r}}$

c) $\hat{\boldsymbol{\theta}}$

d) $-\hat{\boldsymbol{\theta}}$

3. Suppose a circular disk of radius 1 cm has mass density $10\,\mathrm{g/cm^2}$ at its center, and $1\,\mathrm{g/cm^2}$ at its edge, and its density is a linear function of the distance from the center. The total mass of the disk is equal to

a) 8.80 g

b) 10.21 g

c) 12.57 g

d) 17.23 g

Lecture 32 | Cylindrical coordinates

Cylindrical coordinates extends polar coordinates to three dimensions by adding a Cartesian coordinate along the z-axis (see figure). To conform to standard usage, we change notation and define the relationship between the Cartesian and the cylindrical coordinates to be

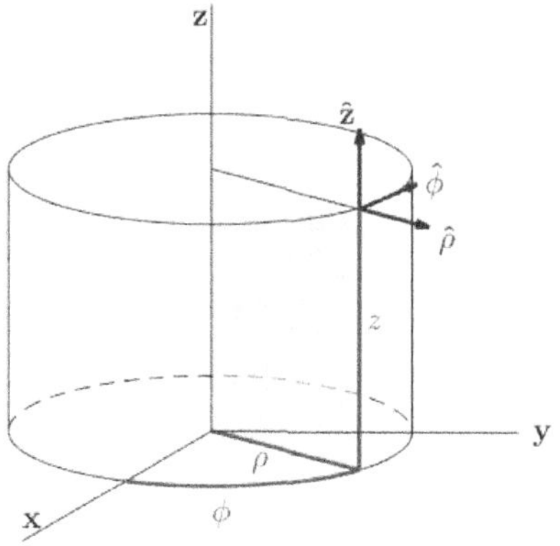

$$x = \rho \cos\phi, \qquad y = \rho \sin\phi, \qquad z = z;$$

with inverse relation

$$\rho = \sqrt{x^2 + y^2}, \qquad \tan\phi = y/x.$$

A spatial point (x, y, z) in Cartesian coordinates is now specified by (ρ, ϕ, z) in cylindrical coordinates.

The orthogonal unit vectors $\hat{\boldsymbol{\rho}}$, $\hat{\boldsymbol{\phi}}$, and $\hat{\boldsymbol{z}}$ point in the direction of increasing ρ, ϕ and z, respectively, and $\hat{\boldsymbol{\rho}}$ and $\hat{\boldsymbol{\phi}}$ are functions of the angle ϕ. The position vector is given by $\boldsymbol{r} = \rho\hat{\boldsymbol{\rho}} + z\hat{\boldsymbol{z}}$. The differential volume element transforms as $dx\,dy\,dz = \rho\,d\rho\,d\phi\,dz$.

The del operator can be found using the polar form of the Cartesian derivatives (see the problems). The result is

$$\boldsymbol{\nabla} = \hat{\boldsymbol{\rho}}\frac{\partial}{\partial\rho} + \hat{\boldsymbol{\phi}}\frac{1}{\rho}\frac{\partial}{\partial\phi} + \hat{\boldsymbol{z}}\frac{\partial}{\partial z}.$$

The Laplacian, $\boldsymbol{\nabla}\cdot\boldsymbol{\nabla}$, is computed taking care to differentiate the unit vectors with respect to ϕ:

$$\begin{aligned}\nabla^2 &= \frac{\partial^2}{\partial\rho^2} + \frac{1}{\rho}\frac{\partial}{\partial\rho} + \frac{1}{\rho^2}\frac{\partial^2}{\partial\phi^2} + \frac{\partial^2}{\partial z^2}\\ &= \frac{1}{\rho}\frac{\partial}{\partial\rho}\left(\rho\frac{\partial}{\partial\rho}\right) + \frac{1}{\rho^2}\frac{\partial^2}{\partial\phi^2} + \frac{\partial^2}{\partial z^2}.\end{aligned}$$

The divergence and curl of a vector field, $\boldsymbol{A} = A_\rho(\rho,\phi,z)\hat{\boldsymbol{\rho}} + A_\phi(\rho,\phi,z)\hat{\boldsymbol{\phi}} + A_z(\rho,\phi,z)\hat{\boldsymbol{z}}$, are given by

$$\boldsymbol{\nabla}\cdot\boldsymbol{A} = \frac{1}{\rho}\frac{\partial}{\partial\rho}(\rho A_\rho) + \frac{1}{\rho}\frac{\partial A_\phi}{\partial\phi} + \frac{\partial A_z}{\partial z},$$

$$\boldsymbol{\nabla}\times\boldsymbol{A} = \hat{\boldsymbol{\rho}}\left(\frac{1}{\rho}\frac{\partial A_z}{\partial\phi} - \frac{\partial A_\phi}{\partial z}\right) + \hat{\boldsymbol{\phi}}\left(\frac{\partial A_\rho}{\partial z} - \frac{\partial A_z}{\partial\rho}\right) + \hat{\boldsymbol{z}}\left(\frac{1}{\rho}\frac{\partial}{\partial\rho}(\rho A_\phi) - \frac{1}{\rho}\frac{\partial A_\rho}{\partial\phi}\right).$$

Problems for Lecture 32

1. Determine the del operator $\boldsymbol{\nabla}$ in cylindrical coordinates. There are several ways to do this, but a straightforward, though algebraically lengthy one, is to transform from Cartesian coordinates using

$$\boldsymbol{\nabla} = \hat{\boldsymbol{x}}\frac{\partial}{\partial x} + \hat{\boldsymbol{y}}\frac{\partial}{\partial y} + \hat{\boldsymbol{z}}\frac{\partial}{\partial z},$$

and the identities

$$\hat{\boldsymbol{x}} = \cos\phi\hat{\boldsymbol{\rho}} - \sin\phi\hat{\boldsymbol{\phi}}, \qquad \hat{\boldsymbol{y}} = \sin\phi\hat{\boldsymbol{\rho}} + \cos\phi\hat{\boldsymbol{\phi}},$$

and

$$\frac{\partial}{\partial x} = \cos\phi\frac{\partial}{\partial\rho} - \frac{\sin\phi}{\rho}\frac{\partial}{\partial\phi}, \qquad \frac{\partial}{\partial y} = \sin\phi\frac{\partial}{\partial\rho} + \frac{\cos\phi}{\rho}\frac{\partial}{\partial\phi}.$$

2. Compute $\boldsymbol{\nabla}\cdot\hat{\boldsymbol{\rho}}$ in two ways:

a) With the divergence in cylindrical coordinates;

b) By transforming to Cartesian coordinates.

3. Using cylindrical coordinates, compute $\boldsymbol{\nabla}\times\hat{\boldsymbol{\rho}}$, $\boldsymbol{\nabla}\cdot\hat{\boldsymbol{\phi}}$ and $\boldsymbol{\nabla}\times\hat{\boldsymbol{\phi}}$.

4. The center-of-mass of a solid with density ρ and total mass M is defined (with respect to a given coordinate system with position vector $\boldsymbol{r}$) as

$$\boldsymbol{R} = \frac{1}{M}\int_V \rho\boldsymbol{r}\,dV.$$

Find the center-of-mass of the uniform solid cone pictured below, with coordinate system specified. You may assume that the volume of the cone is given by $V = \frac{1}{3}\pi a^2 b$.

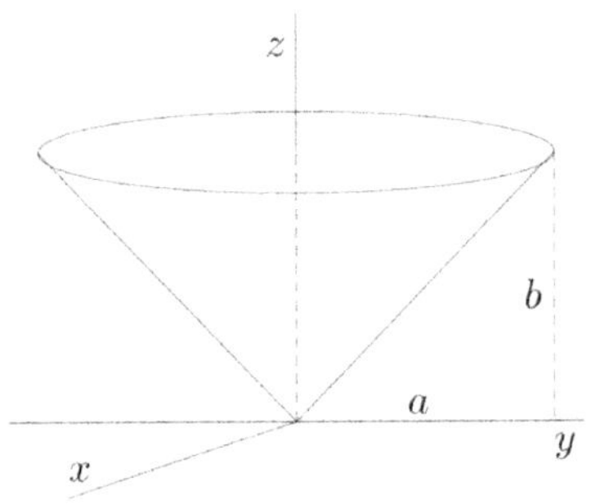

Lecture 33 Spherical coordinates (Part A)

Spherical coordinates are useful for problems with spherical symmetry. The relationship between the Cartesian and the spherical coordinates (see figure) is given by

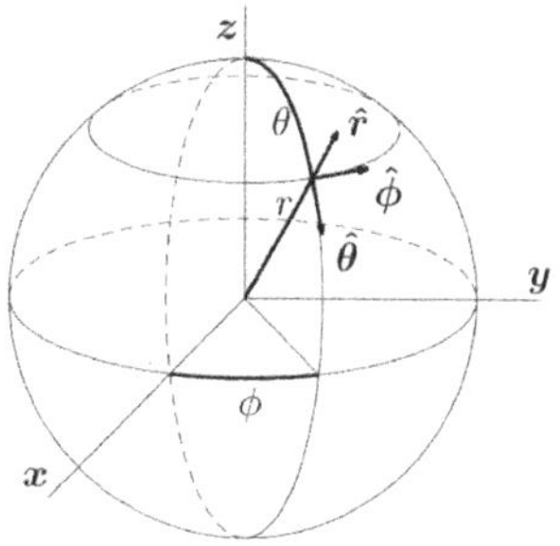

$$x = r\sin\theta\cos\phi, \quad y = r\sin\theta\sin\phi, \quad z = r\cos\theta,$$

where $0 \leq \theta \leq \pi$ and $0 \leq \phi \leq 2\pi$. A spatial point (x, y, z) in Cartesian coordinates is now specified by (r, θ, ϕ) in spherical coordinates. The orthogonal unit vectors $\hat{\boldsymbol{r}}$, $\hat{\boldsymbol{\theta}}$, and $\hat{\boldsymbol{\phi}}$ point in the direction of increasing r, θ and ϕ, respectively.

The position vector is given by

$$\boldsymbol{r} = r\hat{\boldsymbol{r}},$$

where $\hat{\boldsymbol{r}} = \hat{\boldsymbol{r}}(\theta, \phi)$; and the differential volume element transforms as

$$dx\,dy\,dz = r^2 \sin\theta\,dr\,d\theta\,d\phi.$$

Using trigonometry, the spherical coordinate unit vectors can be written in terms of the Cartesian unit vectors by

$$\begin{aligned}
\hat{\boldsymbol{r}} &= \sin\theta\cos\phi\,\boldsymbol{i} + \sin\theta\sin\phi\,\boldsymbol{j} + \cos\theta\,\boldsymbol{k},\\
\hat{\boldsymbol{\theta}} &= \cos\theta\cos\phi\,\boldsymbol{i} + \cos\theta\sin\phi\,\boldsymbol{j} - \sin\theta\,\boldsymbol{k},\\
\hat{\boldsymbol{\phi}} &= -\sin\phi\,\boldsymbol{i} + \cos\phi\,\boldsymbol{j};
\end{aligned}$$

with inverse relation

$$\begin{aligned}
\boldsymbol{i} &= \sin\theta\cos\phi\,\hat{\boldsymbol{r}} + \cos\theta\cos\phi\,\hat{\boldsymbol{\theta}} - \sin\phi\,\hat{\boldsymbol{\phi}},\\
\boldsymbol{j} &= \sin\theta\sin\phi\,\hat{\boldsymbol{r}} + \cos\theta\sin\phi\,\hat{\boldsymbol{\theta}} + \cos\phi\,\hat{\boldsymbol{\phi}},\\
\boldsymbol{k} &= \cos\theta\,\hat{\boldsymbol{r}} - \sin\theta\,\hat{\boldsymbol{\theta}}.
\end{aligned}$$

By differentiating the unit vectors, we can derive the sometimes useful identities

$$\begin{aligned}
\frac{\partial\hat{\boldsymbol{r}}}{\partial\theta} &= \hat{\boldsymbol{\theta}}, & \frac{\partial\hat{\boldsymbol{\theta}}}{\partial\theta} &= -\hat{\boldsymbol{r}}, & \frac{\partial\hat{\boldsymbol{\phi}}}{\partial\theta} &= 0;\\
\frac{\partial\hat{\boldsymbol{r}}}{\partial\phi} &= \sin\theta\hat{\boldsymbol{\phi}}, & \frac{\partial\hat{\boldsymbol{\theta}}}{\partial\phi} &= \cos\theta\hat{\boldsymbol{\phi}}, & \frac{\partial\hat{\boldsymbol{\phi}}}{\partial\phi} &= -\sin\theta\hat{\boldsymbol{r}} - \cos\theta\hat{\boldsymbol{\theta}}.
\end{aligned}$$

Problems for Lecture 33

1. Write the relationship between the spherical coordinate unit vectors and the Cartesian unit vectors in matrix form. Notice that Q, the transformation matrix for this relationship, is orthogonal. Invert the relationship using the formula $Q^{-1} = Q^T$.

2. Use the Jacobian change-of-variables formula for triple integrals, given by

$$dx\,dy\,dz = \left| \det \begin{pmatrix} \partial x/\partial r & \partial x/\partial \theta & \partial x/\partial \phi \\ \partial y/\partial r & \partial y/\partial \theta & \partial y/\partial \phi \\ \partial z/\partial r & \partial z/\partial \theta & \partial z/\partial \phi \end{pmatrix} \right| dr\,d\theta\,d\phi,$$

to derive $dx\,dy\,dz = r^2 \sin\theta\,dr\,d\theta\,d\phi$.

3. Consider a scalar field $f = f(r)$ that depends only on the distance from the origin. Using $dx\,dy\,dz = r^2 \sin\theta\,dr\,d\theta\,d\phi$, and an integration region V inside a sphere of radius R centered at the origin, show that

$$\int_V f\,dV = 4\pi \int_0^R r^2 f(r)\,dr.$$

4. Suppose a sphere of radius R has mass density ρ_0 at its center, and ρ_1 at its surface, and its density is a linear function of the distance from the center. Find the total mass of the sphere. What is the average density of the sphere?

Lecture 34 Spherical coordinates (Part B)

First, we determine the gradient in spherical coordinates. Consider the scalar field $f = f(r, \theta, \phi)$. Our definition of a total differential is

$$df = \frac{\partial f}{\partial r} dr + \frac{\partial f}{\partial \theta} d\theta + \frac{\partial f}{\partial \phi} d\phi = \boldsymbol{\nabla} f \cdot d\boldsymbol{r}.$$

In spherical coordinates,

$$\boldsymbol{r} = r\,\hat{\boldsymbol{r}}(\theta, \phi),$$

and using

$$\frac{\partial \hat{\boldsymbol{r}}}{\partial \theta} = \hat{\boldsymbol{\theta}}, \qquad \frac{\partial \hat{\boldsymbol{r}}}{\partial \phi} = \sin\theta\,\hat{\boldsymbol{\phi}},$$

we have

$$d\boldsymbol{r} = dr\,\hat{\boldsymbol{r}} + r\frac{\partial \hat{\boldsymbol{r}}}{\partial \theta} d\theta + r\frac{\partial \hat{\boldsymbol{r}}}{\partial \phi} d\phi = dr\,\hat{\boldsymbol{r}} + r\,d\theta\,\hat{\boldsymbol{\theta}} + r\sin\theta\,d\phi\,\hat{\boldsymbol{\phi}}.$$

Using the orthonormality of the unit vectors, we can therefore write df as

$$df = \left(\frac{\partial f}{\partial r}\hat{\boldsymbol{r}} + \frac{1}{r}\frac{\partial f}{\partial \theta}\hat{\boldsymbol{\theta}} + \frac{1}{r\sin\theta}\frac{\partial f}{\partial \phi}\hat{\boldsymbol{\phi}}\right) \cdot (dr\,\hat{\boldsymbol{r}} + r\,d\theta\,\hat{\boldsymbol{\theta}} + r\sin\theta\,d\phi\,\hat{\boldsymbol{\phi}}),$$

showing that

$$\boldsymbol{\nabla} f = \frac{\partial f}{\partial r}\hat{\boldsymbol{r}} + \frac{1}{r}\frac{\partial f}{\partial \theta}\hat{\boldsymbol{\theta}} + \frac{1}{r\sin\theta}\frac{\partial f}{\partial \phi}\hat{\boldsymbol{\phi}}.$$

Some messy algebra will yield for the Laplacian of the scalar field f,

$$\nabla^2 f = \frac{1}{r^2}\frac{\partial}{\partial r}\left(r^2\frac{\partial f}{\partial r}\right) + \frac{1}{r^2\sin\theta}\frac{\partial}{\partial \theta}\left(\sin\theta\frac{\partial f}{\partial \theta}\right) + \frac{1}{r^2\sin^2\theta}\frac{\partial^2 f}{\partial \phi^2};$$

and for the divergence and curl of a vector field,

$$\boldsymbol{A} = A_r(r,\theta,\phi)\hat{\boldsymbol{r}} + A_\theta(r,\theta,\phi)\hat{\boldsymbol{\theta}} + A_\phi(r,\theta,\phi)\hat{\boldsymbol{\phi}},$$

$$\boldsymbol{\nabla} \cdot \boldsymbol{A} = \frac{1}{r^2}\frac{\partial}{\partial r}(r^2 A_r) + \frac{1}{r\sin\theta}\frac{\partial}{\partial \theta}(\sin\theta A_\theta) + \frac{1}{r\sin\theta}\frac{\partial A_\phi}{\partial \phi},$$

$$\boldsymbol{\nabla} \times \boldsymbol{A} = \frac{\hat{\boldsymbol{r}}}{r\sin\theta}\left[\frac{\partial}{\partial \theta}(\sin\theta A_\phi) - \frac{\partial A_\theta}{\partial \phi}\right] + \frac{\hat{\boldsymbol{\theta}}}{r}\left[\frac{1}{\sin\theta}\frac{\partial A_r}{\partial \phi} - \frac{\partial}{\partial r}(rA_\phi)\right] + \frac{\hat{\boldsymbol{\phi}}}{r}\left[\frac{\partial}{\partial r}(rA_\theta) - \frac{\partial A_r}{\partial \theta}\right].$$

Problems for Lecture 34

1. Using the formulas for the spherical coordinate unit vectors in terms of the Cartesian unit vectors, prove that

$$\frac{\partial \hat{\boldsymbol{r}}}{\partial \theta} = \hat{\boldsymbol{\theta}}, \qquad \frac{\partial \hat{\boldsymbol{r}}}{\partial \phi} = \sin\theta\, \hat{\boldsymbol{\phi}}.$$

2. Compute the divergence and curl of the spherical coordinate unit vectors.

3. It may be useful to know that the Laplacian in spherical coordinates depends on whether it acts on a a scalar or a vector field. The Laplacian of a vector field in spherical coordinates may be derived using the curl of a curl identity,

$$\nabla^2 \boldsymbol{A} = \boldsymbol{\nabla}(\boldsymbol{\nabla} \cdot \boldsymbol{A}) - \boldsymbol{\nabla} \times (\boldsymbol{\nabla} \times \boldsymbol{A}).$$

We consider here two simple cases. (The general form for the Laplacian in spherical coordinates for a vector field can be found online.)

a) If in spherical coordinates, $f = f(r)$, show that

$$\nabla^2 f = \frac{1}{r^2}\frac{d}{dr}\left(r^2 \frac{df}{dr}\right).$$

b) If in spherical coordinates, $\boldsymbol{A} = A_r(r)\hat{\boldsymbol{r}}$, show that

$$\nabla^2 \boldsymbol{A} = \frac{d}{dr}\left(\frac{1}{r^2}\frac{d}{dr}(r^2 A_r)\right)\hat{\boldsymbol{r}}.$$

4. Using spherical coordinates, show that for $r \neq 0$,

$$\nabla^2\left(\frac{1}{r}\right) = 0.$$

5. Using spherical coordinates, show that for $r \neq 0$,

$$\nabla^2\left(\frac{\hat{\boldsymbol{r}}}{r^2}\right) = 0.$$

Practice Quiz | Cylindrical and spherical coordinates

1. With $\rho = \sqrt{x^2 + y^2}$, the function $\nabla^2 \left(\frac{1}{\rho} \right)$ is equal to

a) 0

b) $\dfrac{1}{\rho^3}$

c) $\dfrac{2}{\rho^3}$

d) $\dfrac{3}{\rho^3}$

2. Let $\boldsymbol{r} = x\boldsymbol{i}$. Then $(\hat{\boldsymbol{r}}, \hat{\boldsymbol{\theta}}, \hat{\boldsymbol{\phi}})$ is equal to

a) $(\boldsymbol{i}, \boldsymbol{j}, \boldsymbol{k})$

b) $(\boldsymbol{i}, \boldsymbol{k}, \boldsymbol{j})$

c) $(\boldsymbol{i}, -\boldsymbol{k}, \boldsymbol{j})$

d) $(\boldsymbol{i}, \boldsymbol{k}, -\boldsymbol{j})$

3. Suppose a sphere of radius 5 cm has mass density $10\,\text{g/cm}^3$ at its center, and 5g/cm^3 at its surface, and its density is a linear function of the distance from the center. The total mass of the sphere is given by

a) 3927 g

b) 3491 g

c) 3272 g

d) 3142 g

Week IV

Line and surface integrals

This week we learn about line integrals and surface integrals. We learn how to take the line integral of a scalar field and use line integrals to compute arc lengths. We then learn how to take line integrals of vector fields by taking the dot product of the vector field with tangent unit vectors to the curve. Consideration of the line integral of a force field results in the work-energy theorem. Next, we learn how to take the surface integral of a scalar field, and compute the surface areas of a cylinder, cone, sphere and paraboloid. We then learn how to take the surface integral of a vector field by taking the dot product of the vector field with normal unit vectors to the surface. The surface integral of a velocity field is used to define the mass flux of a fluid through the surface.

Lecture 35 | Line integral of a scalar field

Define the line integral of a scalar field $f = f(\boldsymbol{r})$ over a curve C by subdividing the curve into small scalar elements of length ds, multiplying each ds by the average value of f on the element, and summing over all elements. We write the line integral as

$$\int_C f(\boldsymbol{r})\, ds.$$

For simplicity, we restrict our discussion here to curves in the x-y plane and write $f(\boldsymbol{r}) = f(x, y)$. By the Pythagorean theorem, we can also write

$$ds = \sqrt{(dx)^2 + (dy)^2},$$

and it is possible to convert a line integral into an ordinary one-dimensional integral in one of two ways. First, if the curve can be specified by a one-dimensional function, such as $y = y(x)$ from $x = x_0$ to x_f, then $dy = y'(x)dx$ and

$$ds = \sqrt{1 + y'(x)^2}\, dx.$$

We then have

$$\int_C f(x, y)\, ds = \int_{x_0}^{x_f} f(x, y(x))\, \sqrt{1 + y'(x)^2}\, dx.$$

Second, if the curve can be parameterized by t, such as $x = x(t)$ and $y = y(t)$, with t ranging from $t = t_0$ to t_f, then $dx = \dot{x}(t)dt$ and $dy = \dot{y}(t)dt$, and

$$ds = \sqrt{\dot{x}(t)^2 + \dot{y}(t)^2}\, dt.$$

The line integral then becomes

$$\int_C f(x, y)\, ds = \int_{t_0}^{t_f} f(x(t), y(t))\, \sqrt{\dot{x}(t)^2 + \dot{y}(t)^2}\, dt.$$

Problems for Lecture 35

1. The perimeter (or circumference) of a closed curve C can be computed from

$$P = \int_C ds.$$

Compute the circumference of a circle of radius R by parameterizing the circle and performing a line integral.

2. The linear mass density of a wire lying flat in the x-y plane can be specified by $\lambda = \lambda(x, y)$, with units of mass per unit length. The total mass of the wire is found from the line integral

$$M = \int_C \lambda(x, y)\, ds,$$

where the integration is over the curve C formed by the wire. Suppose a semi-circular wire of radius R has linear mass density λ_0 on one end and λ_1 on the other end, and its linear mass density increases linearly along the length of the wire. Find the total mass of the wire.

Lecture 36 | Arc length

The arc length P of a curve C, which is a special case of a line integral of a scalar field, is given by

$$P = \int_C ds.$$

For even simple curves, this integral may not have an analytical solution. For example, consider the circumference of an ellipse. The equation for an ellipse centered at the origin with width $2a$ and height $2b$ is given by

$$\frac{x^2}{a^2} + \frac{y^2}{b^2} = 1.$$

Assuming $a \geq b$, the eccentricity of this ellipse is defined as

$$e = \sqrt{1 - (b/a)^2}.$$

To compute the circumference of an ellipse, we parameterize the ellipse using

$$x(\theta) = a\cos\theta, \qquad y(\theta) = b\sin\theta,$$

where θ goes from 0 to 2π. The infinitesimal arc length ds is then given by

$$\begin{aligned}
ds &= \sqrt{(dx)^2 + (dy)^2} = \sqrt{x'(\theta)^2 + y'(\theta)^2}\, d\theta \\
&= \sqrt{a^2\sin^2\theta + b^2\cos^2\theta}\, d\theta \\
&= a\sqrt{1 - [1 - (b/a)^2]\cos^2\theta}\, d\theta \\
&= a\sqrt{1 - e^2\cos^2\theta}\, d\theta.
\end{aligned}$$

Therefore, the circumference of an ellipse — or perimeter P — is given by the line integral

$$\begin{aligned}
P = \int_C ds &= a\int_0^{2\pi} \sqrt{1 - e^2\cos^2\theta}d\theta \\
&= 4a\int_0^{\pi/2} \sqrt{1 - e^2\cos^2\theta}d\theta,
\end{aligned}$$

where the final integral is over one-quarter of the arc length of the ellipse , and is called the complete elliptic integral of the second kind.

Problems for Lecture 36

1. Determine a formula for the perimeter of an elipse that deviates slightly from a circle. Assume that the ellipse is specified by a and b (and the eccentricity e) as defined in the lecture, and that the radius of the circle is given by a. Start with the exact integral formula for the perimeter of an ellipse. Taylor series expand in e keeping only terms up to e^2, and integrate.

Lecture 37 Line integral of a vector field

Let $\boldsymbol{u} = \boldsymbol{u}(\boldsymbol{r})$ be a vector field, C a directed curve, and $d\boldsymbol{r}$ the infinitesimal displacement vector along C. Define a unit vector $\hat{\boldsymbol{t}}$ that points in the direction of $d\boldsymbol{r}$ such that $d\boldsymbol{r} = \hat{\boldsymbol{t}}ds$. Then the line integral of $\boldsymbol{u}$ along C is defined to be

$$\int_C \boldsymbol{u} \cdot d\boldsymbol{r} = \int_C \boldsymbol{u} \cdot \hat{\boldsymbol{t}}\, ds,$$

which is a line integral of the scalar field $\boldsymbol{u} \cdot \hat{\boldsymbol{t}}$. If the curve is closed, sometimes a circle is written in the middle of the integral sign.

A general method to calculate the line integral of a vector field is to parameterize the curve. Let the curve be parameterized by the function $\boldsymbol{r} = \boldsymbol{r}(t)$ as t goes from t_0 to t_f. Using $d\boldsymbol{r} = (d\boldsymbol{r}/dt)dt$, the line integral becomes

$$\int_C \boldsymbol{u} \cdot d\boldsymbol{r} = \int_{t_0}^{t_f} \boldsymbol{u}(\boldsymbol{r}(t)) \cdot \frac{d\boldsymbol{r}}{dt}\, dt.$$

Sometimes the curve is simple enough that $d\boldsymbol{r}$ can be computed directly.

Example: Compute the line integral of $\boldsymbol{r} = x\boldsymbol{i} + y\boldsymbol{j}$ in the x-y plane along two curves from the origin to the point $(x, y) = (1, 1)$. The first curve C_1 consists of two line segments, the first from $(0, 0)$ to $(1, 0)$, and the second from $(1, 0)$ to $(1, 1)$. The second curve C_2 is a straight line from the origin to $(1, 1)$.

The computation along the first curve C_1 requires two separate integrations. For the curve along the x-axis, we use $d\boldsymbol{r} = dx\boldsymbol{i}$ and for the curve at $x = 1$ in the direction of $\boldsymbol{j}$, we use $d\boldsymbol{r} = dy\boldsymbol{j}$. The line integral is therefore given by

$$\int_{C_1} \boldsymbol{r} \cdot d\boldsymbol{r} = \int_0^1 x\, dx + \int_0^1 y\, dy = 1.$$

For the second curve C_2, we parameterize the straight line by $\boldsymbol{r}(t) = t(\boldsymbol{i} + \boldsymbol{j})$ as t goes from 0 to 1, so that $d\boldsymbol{r} = dt(\boldsymbol{i} + \boldsymbol{j})$, and the integral becomes

$$\int_{C_2} \boldsymbol{r} \cdot d\boldsymbol{r} = \int_0^1 2t\, dt = 1.$$

The two line integrals are equal, and for this case depend only on the starting and ending points of the curves.

Problems for Lecture 37

1. In the x-y plane, calculate the line integral of the vector field $\boldsymbol{u} = -y\boldsymbol{i} + x\boldsymbol{j}$ counterclockwise around a square with vertices $(0,0)$, $(L,0)$, (L,L), and $(0,L)$.

2. In the x-y plane, calculate the line integral of the vector field $\boldsymbol{u} = -y\boldsymbol{i} + x\boldsymbol{j}$ counterclockwise around the circle of radius R centered at the origin.

Lecture 38 | Work-energy theorem

Newton's second law of motion for a mass m with velocity $\boldsymbol{v}$ acted on by a force $\boldsymbol{F}$ is given by

$$m\frac{d\boldsymbol{v}}{dt} = \boldsymbol{F}.$$

The conservation of energy is an important concept in physics and requires a definition of work. We first take the dot product of both sides of Newton's law with the velocity vector and use the identity

$$m\boldsymbol{v}\cdot\frac{d\boldsymbol{v}}{dt} = \frac{d}{dt}\left(\frac{1}{2}m|\boldsymbol{v}|^2\right)$$

to obtain

$$\frac{d}{dt}\left(\frac{1}{2}m|\boldsymbol{v}|^2\right) = \boldsymbol{F}\cdot\boldsymbol{v}.$$

Integrating from an initial time t_i to a final time t_f, with $\boldsymbol{v}(t_i) = \boldsymbol{v}_i$ and $\boldsymbol{v}(t_f) = \boldsymbol{v}_f$, we have

$$\frac{1}{2}m|\boldsymbol{v}_f|^2 - \frac{1}{2}m|\boldsymbol{v}_i|^2 = \int_{t_i}^{t_f}\boldsymbol{F}\cdot\boldsymbol{v}\,dt = \int_C \boldsymbol{F}\cdot d\boldsymbol{r},$$

where we have used $d\boldsymbol{r} = \boldsymbol{v}dt$. The line integral is taken along the curve C traversed by the mass between the times t_i and t_f. We define the kinetic energy T of a mass m by

$$T = \frac{1}{2}m|\boldsymbol{v}|^2,$$

and the work W done by a force on a mass as it moves along a curve C as

$$W = \int_C \boldsymbol{F}\cdot d\boldsymbol{r}.$$

We these definitions, the work-energy theorem states that the work done on a mass by a force is equal to the change in the kinetic energy of the mass, or in equation form,

$$W = T_f - T_i.$$

Problems for Lecture 38

1. A mass m is dropped from a building of height h. Assuming the mass falls with constant acceleration $-g$, calculate the work done by gravity and compute the velocity of the mass as it hits the ground.

Practice Quiz | Line integrals

1. The arc length of the parabola $y = x^2$ for $0 \le x \le 1$ is given by the integral

a) $\int_0^1 \sqrt{1+2x}\,dx$

b) $\int_0^1 \sqrt{1+4x}\,dx$

c) $\int_0^1 \sqrt{1+2x^2}\,dx$

d) $\int_0^1 \sqrt{1+4x^2}\,dx$

2. The line integral of the vector field $\boldsymbol{u} = -y\boldsymbol{i} + x\boldsymbol{j}$ counterclockwise around a triangle with vertices $(0,0)$, $(L,0)$, and $(0,L)$ is equal to

a) 0

b) $\frac{1}{2}L^2$

c) L^2

d) $2L^2$

3. A mass m is shot upward from the ground with a speed v_0, attains a maximum height, and then falls back to ground. Calculate the work done by gravity on the mass.

a) 0

b) $v_0^2/4g$

c) $v_0^2/2g$

d) v_0^2/g

Lecture 39 Surface integral of a scalar field

Let S be a surface in a three-dimensional space. Define the surface integral of a scalar field $f = f(\boldsymbol{r})$ over S by subdividing the surface into small elements dS, multiplying each element by the average value of f on the element, and summing over all elements. We write the surface integral as

$$\int_S f(\boldsymbol{r})\, dS.$$

A parameterization of the surface S is given by

$$\boldsymbol{r}(u,v) = x(u,v)\boldsymbol{i} + y(u,v)\boldsymbol{j} + z(u,v)\boldsymbol{k},$$

where the position vector $\boldsymbol{r}$ points to the surface, and the surface is spanned by $\boldsymbol{r}$ as the parameters u and v vary. The surface integral is performed over u and v, and the infinitesimal surface element dS is found from the area defined by the tangent vectors $\partial\boldsymbol{r}/\partial u$ and $\partial\boldsymbol{r}/\partial v$ to the surface by way of the cross product, that is,

$$dS = \left|\frac{\partial \boldsymbol{r}}{\partial u} \times \frac{\partial \boldsymbol{r}}{\partial v}\right| du\, dv.$$

If the surface lies over an area A in the x-y plane and can be describe by $z = z(x,y)$, then we can parameterize the surface by $\boldsymbol{r}(x,y)$ and write

$$\boldsymbol{r}(x,y) = x\boldsymbol{i} + y\boldsymbol{j} + z(x,y)\boldsymbol{k}.$$

Therefore,

$$\frac{\partial \boldsymbol{r}}{\partial x} = \boldsymbol{i} + \frac{\partial z}{\partial x}\boldsymbol{k}, \qquad \frac{\partial \boldsymbol{r}}{\partial y} = \boldsymbol{j} + \frac{\partial z}{\partial y}\boldsymbol{k},$$

and the cross product is given by

$$\frac{\partial \boldsymbol{r}}{\partial x} \times \frac{\partial \boldsymbol{r}}{\partial y} = \begin{vmatrix} \boldsymbol{i} & \boldsymbol{j} & \boldsymbol{k} \\ 1 & 0 & \partial z/\partial x \\ 0 & 1 & \partial z/\partial y \end{vmatrix} = -\frac{\partial z}{\partial x}\boldsymbol{i} - \frac{\partial z}{\partial y}\boldsymbol{j} + \boldsymbol{k}.$$

The surface integral in this case becomes

$$\int_S f(\mathbf{r})\, dS = \int\int_A f(\boldsymbol{r}(x,y))\sqrt{1 + \left(\frac{\partial z}{\partial x}\right)^2 + \left(\frac{\partial z}{\partial y}\right)^2}\, dx\, dy.$$

Problems for Lecture 39

1. Compute the lateral surface area $A = \int_S dS$ of a cylinder (see figure) in two ways.

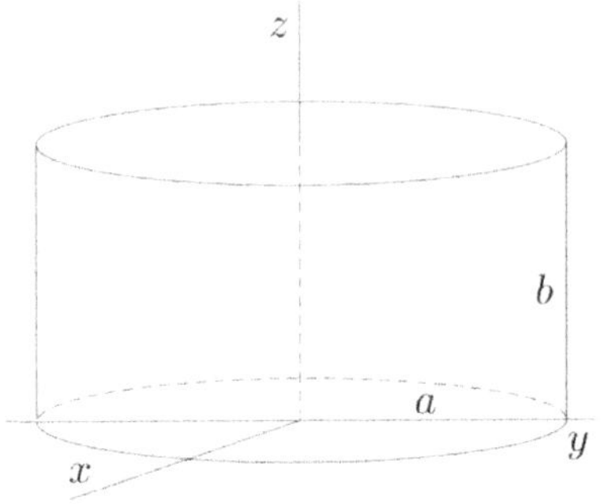

a) Unroll the cylinder and compute the area of the resulting rectangle.

b) Define the cylinder parametrically as

$$\boldsymbol{r} = a\cos\theta\,\boldsymbol{i} + a\sin\theta\,\boldsymbol{j} + z\,\boldsymbol{k}, \qquad \text{for } 0 \le z \le b \quad \text{and} \quad 0 \le \theta \le 2\pi,$$

and compute the surface integral.

2. Compute the lateral surface area $A = \int_S dS$ of a cone (see figure) in two ways.

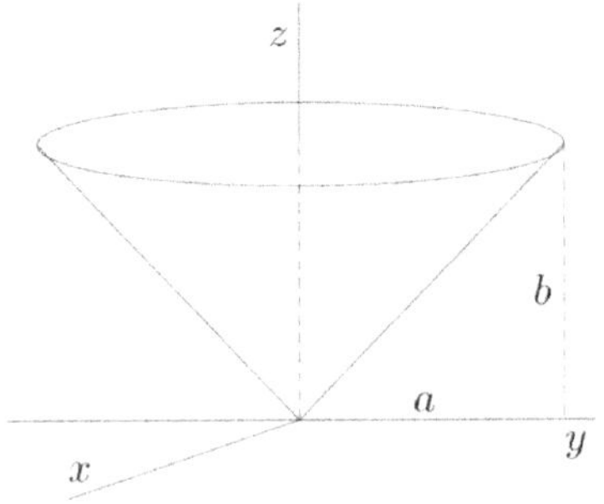

a) Unroll the cone and compute the area of the resulting circular sector.

b) Define the cone parametrically as

$$\boldsymbol{r} = \frac{az}{b}\cos\theta\,\boldsymbol{i} + \frac{az}{b}\sin\theta\,\boldsymbol{j} + z\,\boldsymbol{k}, \qquad \text{for } 0 \le z \le b \quad \text{and} \quad 0 \le \theta \le 2\pi,$$

and compute the surface integral.

Lecture 40 | Surface area of a sphere

Surface area is given by the integral

$$S = \int_S dS,$$

where the integral is evaluated by parameterizing the surface. Here, we compute the surface area of a sphere of radius R. The sphere can be parameterized using spherical coordinates as

$$\boldsymbol{r}(\theta,\phi) = R\sin\theta\cos\phi\boldsymbol{i} + R\sin\theta\sin\phi\boldsymbol{j} + R\cos\theta\boldsymbol{k},$$
$$\text{for } 0 \le \theta \le \pi \text{ and } 0 \le \phi \le 2\pi.$$

To find the infinitesimal surface element, we compute the partial derivatives of $\boldsymbol{r}$:

$$\begin{aligned}\frac{\partial \boldsymbol{r}}{\partial \theta} &= R\cos\theta\cos\phi\,\boldsymbol{i} + R\cos\theta\sin\phi\,\boldsymbol{j} - R\sin\theta\,\boldsymbol{k},\\ \frac{\partial \boldsymbol{r}}{\partial \phi} &= -R\sin\theta\sin\phi\,\boldsymbol{i} + R\sin\theta\cos\phi\,\boldsymbol{j}.\end{aligned}$$

The cross product is

$$\begin{aligned}\frac{\partial \boldsymbol{r}}{\partial \theta} \times \frac{\partial \boldsymbol{r}}{\partial \phi} &= \begin{vmatrix} \boldsymbol{i} & \boldsymbol{j} & \boldsymbol{k} \\ R\cos\theta\cos\phi & R\cos\theta\sin\phi & -R\sin\theta \\ -R\sin\theta\sin\phi & R\sin\theta\cos\phi & 0 \end{vmatrix}\\ &= R^2\sin^2\theta\cos\phi\,\boldsymbol{i} + R^2\sin^2\theta\sin\phi\,\boldsymbol{j} + R^2\sin\theta\cos\theta\,\boldsymbol{k},\end{aligned}$$

so that

$$\begin{aligned}\left|\frac{\partial \boldsymbol{r}}{\partial \theta} \times \frac{\partial \boldsymbol{r}}{\partial \phi}\right| &= R^2\sqrt{\sin^4\theta\cos^2\phi + \sin^4\theta\sin^2\phi + \sin^2\theta\cos^2\theta}\\ &= R^2\sin\theta\sqrt{\sin^2\theta(\sin^2\phi + \cos^2\phi) + \cos^2\theta}\\ &= R^2\sin\theta.\end{aligned}$$

Therefore,

$$dS = \left|\frac{\partial \boldsymbol{r}}{\partial \theta} \times \frac{\partial \boldsymbol{r}}{\partial \phi}\right| d\theta\, d\phi = R^2\sin\theta\, d\theta\, d\phi.$$

Recall that for a sphere, $dV = r^2\sin\theta\, dr\, d\theta\, d\phi$, so that as one would expect, $dSdr$ is equal to dV at $r = R$.

The surface area of a sphere can be found from the integral

$$S = \int_S dS = \int_0^{2\pi}\int_0^{\pi} R^2\sin\theta\, d\theta\, d\phi = R^2\int_0^{2\pi} d\phi \int_0^{\pi}\sin\theta\, d\theta = 4\pi R^2.$$

Problems for Lecture 40

1. Compute the lateral surface area $S = \int_S dS$ of a paraboloid (see figure), defined by

$$a^2 z = b\left(x^2 + y^2\right), \qquad \text{for } 0 \leq z \leq b.$$

Compute the resulting two-dimensional integral using polar coordinates.

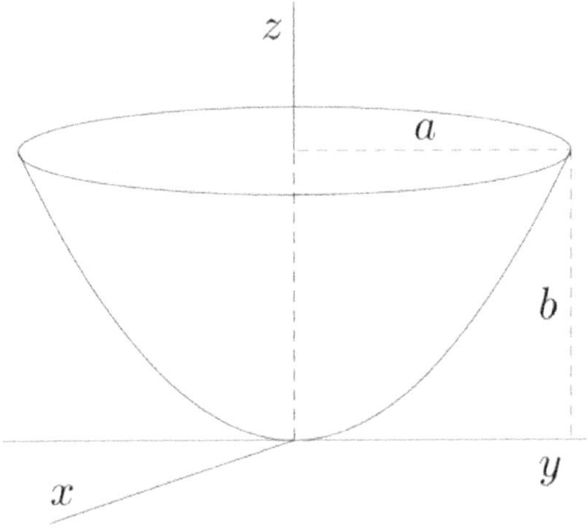

Lecture 41 Surface integral of a vector field

Let $\boldsymbol{u} = \boldsymbol{u}(\boldsymbol{r})$ be a vector field, and let S be a surface with normal unit vector $\hat{\boldsymbol{n}}$. The surface integral of $\boldsymbol{u}$ over S is defined to be

$$\int_S \boldsymbol{u} \cdot d\boldsymbol{S} = \int_S \boldsymbol{u} \cdot \hat{\boldsymbol{n}}\, dS,$$

which is a surface integral of the scalar field $\boldsymbol{u} \cdot \hat{\boldsymbol{n}}$. The integration surface can be either open or closed. For an open surface, the direction of the normal vectors needs to be specified (such as up or down), but for a closed surface, $\hat{\boldsymbol{n}}$ is always assumed to be in the outward direction.

Example: Compute the surface integral of $\boldsymbol{r} = x\boldsymbol{i} + y\boldsymbol{j} + z\boldsymbol{k}$ over a cube centered at the origin with sides parallel to the axes and side lengths equal to L.

The faces of the cube are located at $x = \pm L/2$, $y = \pm L/2$ and $z = \pm L/2$, and the surface area of each face is L^2. For the face at $x = L/2$, say, we have $\boldsymbol{r} = (L/2)\boldsymbol{i} + y\boldsymbol{j} + z\boldsymbol{k}$ and $d\boldsymbol{S} = \boldsymbol{i}dS$. The integral over this face is given by

$$\int_S \boldsymbol{r} \cdot d\boldsymbol{S} = \frac{L}{2}\int_S dS = \frac{L^3}{2}.$$

One can see that the surface integrals over all six faces of the cube are equal, and we obtain for the surface integral over the entire cube,

$$\oint_S \boldsymbol{r} \cdot d\boldsymbol{S} = 6 \times \frac{L^3}{2} = 3L^3,$$

equal to three times the volume of the cube. The circle notation on the integral sign signifies integration over a closed surface.

Example: Compute the surface integral of $\boldsymbol{r} = x\boldsymbol{i} + y\boldsymbol{j} + z\boldsymbol{k}$ over a sphere centered at the origin with radius R.

Using spherical coordinates, on the surface of a sphere of radius R centered at the origin, we have $\boldsymbol{r} = R\hat{\boldsymbol{r}}$, $d\boldsymbol{S} = \hat{\boldsymbol{r}}dS$ and the surface area of the sphere is $4\pi R^2$. Therefore,

$$\oint_S \boldsymbol{r} \cdot d\boldsymbol{S} = R\oint_S dS = 4\pi R^3,$$

equal to three times the volume of the sphere.

Problems for Lecture 41

1. Compute the surface integral of $\boldsymbol{r} = x\boldsymbol{i} + y\boldsymbol{j} + z\boldsymbol{k}$ over a closed cylinder centered at the origin with with radius a and length l.

Lecture 42 Flux integrals

The surface integral of a vector field is often called a flux integral. If $\boldsymbol{u}$ is the fluid velocity (length divided by time), and ρ is the fluid density (mass divided by volume), then the surface integral

$$\int_S \rho \boldsymbol{u} \cdot d\boldsymbol{S}$$

computes the mass flux, that is, the mass passing through the surface S per unit time. Flux integrals for the electric and magnetic vector fields are also defined.

If S is a closed surface, then the normal vector $\hat{\boldsymbol{n}}$ is assumed to be in the outward direction, and a positive value for the flux integral implies a net flux from inside the surface to outside; a negative value implies a net flux from outside to inside. If a fluid is incompressible, a positive mass flux indicates a source of fluid inside the closed surface, and a negative mass flux indicates a sink.

Example: Find the flux of the electric field through a sphere of radius R centered at the origin, where a point charge q is located. The electric field due to the point charge is given in spherical coordinates by Coulomb's law,

$$\boldsymbol{E} = \frac{q}{4\pi\epsilon_0 r^2}\hat{\boldsymbol{r}}.$$

To compute the flux integral, we use $d\boldsymbol{S} = \hat{\boldsymbol{r}}\, dS$, where the surface area of the sphere is given by $S = 4\pi R^2$. We have

$$\oint_S \boldsymbol{E} \cdot d\boldsymbol{S} = \frac{q}{4\pi\epsilon_0 R^2}\oint dS = \frac{q}{\epsilon_0},$$

which is observed to be independent of R because of the Coulomb inverse square law.

Problems for Lecture 42

1. Calculate the mass flux of a laminar fluid of density ρ, viscosity ν and constant pressure gradient G passing through a cross section of a pipe of radius R. Choosing z as the symmetry axis for the pipe, the velocity of the fluid is given by

$$\boldsymbol{u}(\boldsymbol{r}) = u_{\mathrm{m}} \left(1 - \left(\frac{r}{R}\right)^2\right) \boldsymbol{k},$$

where r is the radial coordinate in the cross section and

$$u_{\mathrm{m}} = \frac{GR^2}{4\nu\rho}$$

is the maximum velocity of the fluid in the center of the pipe.

Practice Quiz | Surface integrals

1. The shape of a donut or bagel is called a torus. Let R be the radius from the center of the hole to the center of the torus tube and r be the radius of the torus tube. Then the equation for a torus symmetric about the z-axis is given by

$$\left(R - \sqrt{x^2 + y^2}\right)^2 + z^2 = r^2.$$

The torus may be parameterized by

$$x = (R + r\cos\theta)\cos\phi, \qquad y = (R + r\cos\theta)\sin\phi, \qquad z = r\sin\theta,$$

where $0 \le \theta \le 2\pi$ and $0 \le \phi \le 2\pi$. The infinitesimal surface element dS for the torus is given by

a) $R(r + R\cos\phi)\, d\theta\, d\phi$

b) $r(R + r\cos\phi)\, d\theta\, d\phi$

c) $R(r + R\cos\theta)\, d\theta\, d\phi$

d) $r(R + r\cos\theta)\, d\theta\, d\phi$

2. Consider a closed right circular cylinder of radius R and length L centered on the z-axis. The surface integral of $\boldsymbol{u} = x\boldsymbol{i} + y\boldsymbol{j}$ over the cylinder is given by

a) 0

b) $\pi R^2 L$

c) $2\pi R^2 L$

d) $4\pi R^2 L$

3. The flux integral of $\boldsymbol{u} = z\boldsymbol{k}$ over the upper hemisphere of a sphere of radius R centered at the origin with normal vector $\hat{\boldsymbol{r}}$ is given by

a) $\dfrac{2\pi}{3}R^3$

b) $\dfrac{4\pi}{3}R^3$

c) $2\pi R^3$

d) $4\pi R^3$

Week V

Fundamental Theorems

In this week's lectures, we learn about the fundamental theorems of vector calculus. These include the gradient theorem, the divergence theorem, and Stokes' theorem. We show how these theorems are used to derive the law of conservation of energy, continuity equations, define the divergence and curl in coordinate-free form, and convert the integral version of Maxwell's equations to differential form.

Lecture 43 | Gradient theorem

The gradient theorem is a generalization of the fundamental theorem of calculus for line integrals. Let $\boldsymbol{\nabla}\phi$ be the gradient of a scalar field $\phi = \phi(\boldsymbol{r})$, and let C be a directed curve that begins at the point $\boldsymbol{r}_1$ and ends at $\boldsymbol{r}_2$. Suppose we can parameterize the curve C by $\boldsymbol{r} = \boldsymbol{r}(t)$, where $t_1 \leq t \leq t_2$, $\boldsymbol{r}(t_1) = \boldsymbol{r}_1$, and $\boldsymbol{r}(t_2) = \boldsymbol{r}_2$. Then using the chain rule in the form

$$\frac{d}{dt}\phi(\boldsymbol{r}) = \boldsymbol{\nabla}\phi(\boldsymbol{r}) \cdot \frac{d\boldsymbol{r}}{dt},$$

and the standard fundamental theorem of calculus, we have

$$\begin{aligned}\int_C \boldsymbol{\nabla}\phi \cdot d\boldsymbol{r} &= \int_{t_1}^{t_2} \boldsymbol{\nabla}\phi(\boldsymbol{r}) \cdot \frac{d\boldsymbol{r}}{dt}\,dt = \int_{t_1}^{t_2} \frac{d}{dt}\phi(\boldsymbol{r})\,dt \\ &= \phi(\boldsymbol{r}(t_2)) - \phi(\boldsymbol{r}(t_1)) = \phi(\boldsymbol{r}_2) - \phi(\boldsymbol{r}_1).\end{aligned}$$

A more direct way to derive this result is to write the differential

$$d\phi = \boldsymbol{\nabla}\phi \cdot d\boldsymbol{r},$$

so that

$$\int_C \boldsymbol{\nabla}\phi \cdot d\boldsymbol{r} = \int_C d\phi = \phi(\boldsymbol{r}_2) - \phi(\boldsymbol{r}_1).$$

We have thus shown that the line integral of the gradient of a function is path independent, depending only on the endpoints of the curve. In particular, we have the general result that

$$\oint_C \boldsymbol{\nabla}\phi \cdot d\boldsymbol{r} = 0$$

for any closed curve C.

Example: Compute the line integral of $\boldsymbol{r} = x\boldsymbol{i} + y\boldsymbol{j}$ in the x-y plane from the origin to the point $(1,1)$.

We have $\boldsymbol{r} = \frac{1}{2}\boldsymbol{\nabla}(x^2 + y^2)$, so that the line integral is path independent. Therefore,

$$\int_C \boldsymbol{r} \cdot d\boldsymbol{r} = \frac{1}{2}\int_C \boldsymbol{\nabla}(x^2 + y^2) \cdot d\boldsymbol{r} = \frac{1}{2}(x^2 + y^2)\Big|_{(0,0)}^{(1,1)} = 1.$$

Problems for Lecture 43

1. Let $\phi(\boldsymbol{r}) = x^2y + xy^2 + z$.

   a) Compute $\boldsymbol{\nabla}\phi$.

   b) Compute $\int_C \boldsymbol{\nabla}\phi \cdot d\boldsymbol{r}$ from $(0,0,0)$ to $(1,1,1)$ using the gradient theorem.

   c) Compute $\int_C \boldsymbol{\nabla}\phi \cdot d\boldsymbol{r}$ along the lines segments $(0,0,0)$ to $(1,0,0)$ to $(1,1,0)$ to $(1,1,1)$.

Lecture 44 Conservative vector fields

For a vector field $\boldsymbol{u}$ defined on $\mathcal{R}^3$, except perhaps at isolated singularities, the following conditions are equivalent:

1. $\boldsymbol{\nabla} \times \boldsymbol{u} = 0$;
2. $\boldsymbol{u} = \boldsymbol{\nabla}\phi$ for some scalar field $\phi = \phi(\boldsymbol{r})$;
3. $\int_C \boldsymbol{u} \cdot d\boldsymbol{r}$ is path independent for any curve C;
4. $\oint_C \boldsymbol{u} \cdot d\boldsymbol{r} = 0$ for any closed curve C.

When these conditions hold, we say that $\boldsymbol{u}$ is a conservative vector field.
Example: Let $\boldsymbol{u}(x,y) = x^2(1+y^3)\boldsymbol{i} + y^2(1+x^3)\boldsymbol{j}$. Show that $\boldsymbol{u}$ is a conservative vector field, and determine $\phi = \phi(x,y)$ such that $\boldsymbol{u} = \boldsymbol{\nabla}\phi$.
To show that $\boldsymbol{u}$ is a conservative vector field, we can prove $\boldsymbol{\nabla} \times \boldsymbol{u} = 0$:

$$\boldsymbol{\nabla} \times \boldsymbol{u} = \begin{vmatrix} \boldsymbol{i} & \boldsymbol{j} & \boldsymbol{k} \\ \partial/\partial x & \partial/\partial y & \partial/\partial z \\ x^2(1+y^3) & y^2(1+x^3) & 0 \end{vmatrix} = (3x^2y^2 - 3x^2y^2)\boldsymbol{k} = 0.$$

To find the scalar field ϕ, we solve

$$\frac{\partial \phi}{\partial x} = x^2(1+y^3), \qquad \frac{\partial \phi}{\partial y} = y^2(1+x^3).$$

Integrating the first equation with respect to x holding y fixed, we find

$$\phi = \int x^2(1+y^3)\,dx = \frac{1}{3}x^3(1+y^3) + f(y),$$

where $f = f(y)$ is a function that depends only on y. Differentiating ϕ with respect to y and using the second equation, we obtain

$$x^3y^2 + f'(y) = y^2(1+x^3) \qquad \text{or} \qquad f'(y) = y^2.$$

One more integration results in $f(y) = y^3/3 + c$, with c constant, and the scalar field is given by

$$\phi(x,y) = \frac{1}{3}(x^3 + x^3y^3 + y^3) + c.$$

Problems for Lecture 44

1. Let $\boldsymbol{u} = (2xy + z^2)\boldsymbol{i} + (2yz + x^2)\boldsymbol{j} + (2zx + y^2)\boldsymbol{k}$.

   a) Show that $\boldsymbol{u}$ is a conservative vector field.

   b) Calculate the scalar field ϕ such that $\boldsymbol{u} = \boldsymbol{\nabla}\phi$.

Lecture 45 | Conservation of energy

The work-energy theorem states that the work done on a mass by a force is equal to the change in the kinetic energy of the mass, or

$$\int_C \boldsymbol{F} \cdot d\boldsymbol{r} = T_f - T_i,$$

where the kinetic energy of a mass m moving with velocity $\boldsymbol{v}$ is given by

$$T = \frac{1}{2} m |\boldsymbol{v}|^2.$$

If $\boldsymbol{F}$ is a conservative vector field, then we can write

$$\boldsymbol{F} = -\boldsymbol{\nabla} V,$$

where $V = V(\boldsymbol{r})$ is called the potential energy. Using the gradient theorem, we have

$$T_f - T_i = -\int_C \boldsymbol{\nabla} V \cdot d\boldsymbol{r} = V_i - V_f,$$

where V_i and V_f are the initial and final potential energies of the mass. Rearranging terms, we have

$$T_i + V_i = T_f + V_f.$$

In other words, the sum of the kinetic and potential energy is conserved.

Example: Find the potential energy of mass m in the gravitational field of mass M.

We place the origin of our coordinate system on the mass M. The gravitational force on m at position $\boldsymbol{r}$ is then given by the inverse square law, written as

$$\boldsymbol{F} = -G \frac{mM\boldsymbol{r}}{r^3}.$$

In the problems of Lecture 17, we have shown that $\boldsymbol{\nabla}(1/r) = -\boldsymbol{r}/r^3$. Therefore, the potential energy of m is given by

$$V = -G \frac{mM}{r}.$$

Problems for Lecture 45

1. The escape velocity is the smallest initial velocity for a mass on the Earth's surface to escape from the Earth's gravitational field. Using the conservation of energy, determine the escape velocity of a mass m. Define M to be the mass of the Earth, and R its radius. The gravitational constant G and the acceleration due to gravity on the surface of the Earth g are related by

$$g = \frac{GM}{R^2}.$$

Write the escape velocity in terms of g and R.

Practice Quiz | Gradient theorem

1. Let $\phi(\boldsymbol{r}) = xyz$. The value of $\int_C \nabla\phi \cdot d\boldsymbol{r}$ from $(0,0,0)$ to $(1,1,1)$ is equal to

a) 0

b) 1

c) 2

d) 3

2. Let $\boldsymbol{u} = y\boldsymbol{i} + x\boldsymbol{j}$. The value of $\oint_C \boldsymbol{u} \cdot d\boldsymbol{r}$, where C is the unit circle centered at the origin, is given by

a) 0

b) 1

c) 2

d) 3

3. Let $\boldsymbol{u} = (2x + y)\boldsymbol{i} + (2y + x)\boldsymbol{j} + \boldsymbol{k}$. If $\boldsymbol{u} = \nabla\phi$, then ϕ can be equal to

a) $(x + y)^2 + z$

b) $(x - y)^2 + z$

c) $x^2 + xy + y^2 + z$

d) $x^2 - xy + y^2 + z$

Lecture 46 Divergence theorem

Let $\boldsymbol{u}$ be a differentiable vector field defined inside and on a smooth closed surface S enclosing a volume V. The divergence theorem states that the integral of the divergence of $\boldsymbol{u}$ over the enclosed volume is equal to the flux of $\boldsymbol{u}$ through the bounding surface; that is,

$$\int_V (\boldsymbol{\nabla} \cdot \boldsymbol{u})\, dV = \oint_S \boldsymbol{u} \cdot d\boldsymbol{S}.$$

We first prove the divergence theorem for a rectangular solid with sides parallel to the axes. Let the rectangular solid be defined by $a \leq x \leq b$, $c \leq y \leq d$, and $e \leq z \leq f$. With $\boldsymbol{u} = u_1\boldsymbol{i} + u_2\boldsymbol{j} + u_3\boldsymbol{k}$, the volume integral over V becomes

$$\int_V (\boldsymbol{\nabla} \cdot \boldsymbol{u})\, dV = \int_e^f \int_c^d \int_a^b \left(\frac{\partial u_1}{\partial x} + \frac{\partial u_2}{\partial y} + \frac{\partial u_3}{\partial z} \right) dx\, dy\, dz.$$

The three terms in the integral can be integrated separately using the fundamental theorem of calculus. Each term in succession is integrated as

$$\int_e^f \int_c^d \left(\int_a^b \frac{\partial u_1}{\partial x}\, dx \right) dy\, dz = \int_e^f \int_c^d (u_1(b,y,z) - u_1(a,y,z))\, dy\, dz;$$

$$\int_e^f \int_a^b \left(\int_c^d \frac{\partial u_2}{\partial y}\, dy \right) dx\, dz = \int_e^f \int_a^b (u_2(x,d,z) - u_2(x,c,z))\, dx\, dz;$$

$$\int_c^d \int_a^b \left(\int_e^f \frac{\partial u_3}{\partial z}\, dz \right) dx\, dy = \int_c^d \int_a^b (u_3(x,y,f) - u_3(x,y,e))\, dx\, dy.$$

The integrals on the right-hand-sides correspond exactly to flux integrals over opposite sides of the rectangular solid. For example, the side located at $x = b$ corresponds with $d\boldsymbol{S} = \boldsymbol{i}\, dy\, dz$ and the side located at $x = a$ corresponds with $d\boldsymbol{S} = -\boldsymbol{i}\, dy\, dz$. Summing all three integrals yields the flux of $\mathbf{u}$ through the six-sided bounding surface, thus proving the divergence theorem for a rectangular solid.

Now, given any volume enclosed by a smooth surface, we can subdivide the volume by a very fine three-dimensional rectangular grid and apply the above result to each rectangular solid in the grid. All the volume integrals over the rectangular solids add. The internal rectangular solids, however, share connecting side faces through which the flux integrals cancel, and the only flux integrals that remain are those from the rectangular solids on the boundary of the volume with outward facing surfaces. The result is the divergence theorem for any volume V enclosed by a smooth surface S.

Problems for Lecture 46

1. Prove the divergence theorem for a sphere of radius R centered at the origin. Use spherical coordinates.

Lecture 47 | Divergence theorem (example 1)

The divergence theorem is given by

$$\int_V (\boldsymbol{\nabla} \cdot \boldsymbol{u})\, dV = \oint_S \boldsymbol{u} \cdot d\boldsymbol{S}.$$

Test the divergence theorem using $\boldsymbol{u} = xy\,\boldsymbol{i} + yz\,\boldsymbol{j} + zx\,\boldsymbol{k}$ *for a cube of side L lying in the first octant with a vertex at the origin.*

Here, Cartesian coordinates are appropriate and we use $\boldsymbol{\nabla} \cdot \boldsymbol{u} = y + z + x$. We have for the left-hand side of the divergence theorem,

$$\begin{aligned}\int_V (\boldsymbol{\nabla} \cdot \boldsymbol{u})\, dV &= \int_0^L \int_0^L \int_0^L (x + y + z)\, dx\, dy\, dz \\ &= L^4/2 + L^4/2 + L^4/2 \\ &= 3L^4/2.\end{aligned}$$

For the right-hand side of the divergence theorem, the flux integral only has nonzero contributions from the three sides located at $x = L$, $y = L$ and $z = L$. The corresponding unit normal vectors are $\boldsymbol{i}$, $\boldsymbol{j}$ and $\boldsymbol{k}$, and the corresponding integrals are

$$\begin{aligned}\oint_S \boldsymbol{u} \cdot d\boldsymbol{S} &= \int_0^L \int_0^L Ly\, dy\, dz + \int_0^L \int_0^L Lz\, dx\, dz + \int_0^L \int_0^L Lx\, dx\, dy \\ &= L^4/2 + L^4/2 + L^4/2 \\ &= 3L^4/2.\end{aligned}$$

Problems for Lecture 47

1. Test the divergence theorem using $\boldsymbol{u} = x^2y\,\boldsymbol{i} + y^2z\,\boldsymbol{j} + z^2x\,\boldsymbol{k}$ for a cube of side L lying in the first octant with a vertex at the origin.

2. Compute the flux integral of $\boldsymbol{r} = x\boldsymbol{i} + y\boldsymbol{j} + z\boldsymbol{k}$ over a square box with side lengths equal to L by applying the divergence theorem to convert the flux integral into a volume integral.

Lecture 48 Divergence theorem (example 2)

The divergence theorem is given by

$$\int_V (\boldsymbol{\nabla} \cdot \boldsymbol{u})\, dV = \oint_S \boldsymbol{u} \cdot d\boldsymbol{S}.$$

Test the divergence theorem using $\boldsymbol{u} = r^2\hat{\boldsymbol{r}}$ for a sphere of radius R centered at the origin.

To compute the left-hand-side of the divergence theorem, we recall the formula for the divergence of a vector field $\boldsymbol{u}$ in spherical coordinates:

$$\boldsymbol{\nabla} \cdot \boldsymbol{u} = \frac{1}{r^2}\frac{\partial}{\partial r}(r^2 u_r) + \frac{1}{r\sin\theta}\frac{\partial}{\partial\theta}(\sin\theta u_\theta) + \frac{1}{r\sin\theta}\frac{\partial u_\phi}{\partial\phi}.$$

Here, $u_r = r^2$ is the only nonzero component of $\boldsymbol{u}$, and we have

$$\boldsymbol{\nabla} \cdot \boldsymbol{u} = \frac{1}{r^2}\frac{d}{dr}\left(r^4\right) = 4r.$$

Therefore, using $dV = r^2 \sin\theta\, dr\, d\theta\, d\phi$, we have

$$\begin{aligned}\int_V (\boldsymbol{\nabla} \cdot \boldsymbol{u})\, dV &= \int_0^{2\pi}\int_0^{\pi}\int_0^{R} 4r^3 \sin\theta\, dr\, d\theta\, d\phi \\ &= \int_0^{2\pi} d\phi \int_0^{\pi} \sin\theta\, d\theta \int_0^{R} 4r^3\, dr = 4\pi R^4.\end{aligned}$$

For the right-hand-side of the divergence theorem, we have $\boldsymbol{u} = R^2\hat{\boldsymbol{r}}$ and $d\boldsymbol{S} = \hat{\boldsymbol{r}} R^2 \sin\theta\, d\theta\, d\phi$, so that

$$\oint_S \boldsymbol{u} \cdot d\boldsymbol{S} = \int_0^{2\pi}\int_0^{\pi} R^4 \sin\theta\, d\theta\, d\phi = 4\pi R^4.$$

Problems for Lecture 48

1. Test the divergence theorem using $\boldsymbol{u} = \hat{\boldsymbol{r}}/r$ for a sphere of radius R centered at the origin.

2. Compute the flux integral of $\boldsymbol{r} = x\boldsymbol{i} + y\boldsymbol{j} + z\boldsymbol{k}$ over a sphere of radius R by applying the divergence theorem to convert the flux integral into a volume integral.

3. With Λ a constant, consider the velocity field of a fluid given by

$$\boldsymbol{u}(x, y, z) = \frac{\Lambda(x\boldsymbol{i} + y\boldsymbol{j} + z\boldsymbol{k})}{4\pi(x^2 + y^2 + z^2)^{3/2}}.$$

a) Using spherical coordinates, show that

$$\boldsymbol{u}(r) = \frac{\Lambda\hat{\boldsymbol{r}}}{4\pi r^2}.$$

b) Using spherical coordinates, show that $\boldsymbol{\nabla} \cdot \boldsymbol{u} = 0$ provided $r \neq 0$.

c) Using the divergence theorem, show that

$$\int_V \boldsymbol{\nabla} \cdot \boldsymbol{u}\, dV = \Lambda,$$

provided that the volume V contains the origin, and is zero otherwise. You have therefore shown that the divergence of the velocity field is given by

$$\boldsymbol{\nabla} \cdot \boldsymbol{u} = \Lambda\delta(\boldsymbol{r}),$$

where $\delta(\boldsymbol{r})$ is the three-dimensional Dirac delta function. This velocity field is called a source flow.

Lecture 49 | Continuity equation

The divergence theorem is often used to derive a continuity equation, which expresses the local conservation of some physical quantity such as mass or electric charge. Here, we derive the continuity equation for a compressible fluid such as a gas. Let the scalar function $\rho(\boldsymbol{r}, t)$ be the mass density of a fluid at position $\boldsymbol{r}$ and time t, and $\boldsymbol{u}(\boldsymbol{r}, t)$ be the fluid velocity. We will assume no sources or sinks of fluid. We consider a small test volume V in the fluid flow and consider the change in the fluid mass M inside V.

The fluid mass M in V varys because of the mass flux through the surface S surrounding V, and one has

$$\frac{dM}{dt} = -\oint_S \rho \boldsymbol{u} \cdot d\boldsymbol{S}.$$

Now the mass of the fluid is given in terms of the mass density by

$$M = \int_V \rho \, dV,$$

and application of the divergence theorem to the surface integral results in

$$\frac{d}{dt} \int_V \rho \, dV = -\int_V \boldsymbol{\nabla} \cdot (\rho \boldsymbol{u}) \, dV.$$

Taking the time derivative inside the integral on the left-hand side, and combining the two integrals yields

$$\int_V \left(\frac{\partial \rho}{\partial t} + \boldsymbol{\nabla} \cdot (\rho \boldsymbol{u}) \right) dV = 0.$$

Since this integral vanishes for any test volume placed in the fluid, the integrand itself must be zero everywhere, and we have derived the continuity equation

$$\frac{\partial \rho}{\partial t} + \boldsymbol{\nabla} \cdot (\rho \boldsymbol{u}) = 0.$$

For an incompressible fluid, for which the mass density ρ is uniform and constant, the continuity equation reduces to

$$\boldsymbol{\nabla} \cdot \boldsymbol{u} = 0.$$

A vector field with zero divergence is called incompressible or solenoidal.

Problems for Lecture 49

1. Show that the continuity equation can be written as

$$\frac{\partial \rho}{\partial t} + \boldsymbol{u} \cdot \boldsymbol{\nabla} \rho + \rho \boldsymbol{\nabla} \cdot \boldsymbol{u} = 0.$$

2. The electric charge density (charge per unit volume) is usually written using the same symbol as the mass density, $\rho(\boldsymbol{r}, t)$, and the volume current density (current per unit area) is given by $\boldsymbol{J}(\boldsymbol{r}, t)$. Local conservation of charge states that the time rate of change of the total charge within a volume is equal to the negative of the charge flowing out of that volume, resulting in the equation

$$\frac{d}{dt} \int_V \rho(\boldsymbol{r}, t)\, dV = -\oint_S \boldsymbol{J} \cdot d\boldsymbol{S}.$$

From this law of charge conservation, derive the electrodynamics continuity equation.

Practice Quiz | Divergence theorem

1. The integral of $\boldsymbol{u} = yz\boldsymbol{i} + xz\boldsymbol{j} + xy\boldsymbol{k}$ over the closed surface of a right circular cone with radius R and length L and base in the x-y plane is given by

a) 0

b) $\pi RL\sqrt{R^2 + L^2}$

c) $2\pi RL\sqrt{R^2 + L^2}$

d) $3\pi RL\sqrt{R^2 + L^2}$

2. The surface integral $\oint_S \boldsymbol{r} \cdot d\boldsymbol{S}$ over a right circular cylinder of radius R and length L is equal to

a) 0

b) $\pi R^2 L$

c) $2\pi R^2 L$

d) $3\pi R^2 L$

3. Which velocity field is not incompressible ($\boldsymbol{\nabla} \cdot \boldsymbol{u} \neq 0$)?

a) $\boldsymbol{u} = xy\boldsymbol{i} - \frac{1}{2}y^2\boldsymbol{j}$

b) $\boldsymbol{u} = (1 + x)\boldsymbol{i} + (1 - y)\boldsymbol{j}$

c) $\boldsymbol{u} = (x^2 - xy)\boldsymbol{i} + \left(\frac{1}{2}y^2 - 2xy\right)\boldsymbol{j}$

d) $\boldsymbol{u} = (x + y)^2\boldsymbol{i} + (x - y)^2\boldsymbol{j}$

Lecture 50 Green's theorem

Green's theorem is a two-dimensional version of Stokes' theorem (which applies to three dimensions), and serves as a simpler introduction. Let $\boldsymbol{u} = u_1(x,y)\boldsymbol{i} + u_2(x,y)\boldsymbol{j}$ be a differentiable two-dimensional vector field defined on the x-y plane. Green's theorem relates an area integral over S in the plane to a line integral around C surrounding this area, and is given by

$$\int_S \left(\frac{\partial u_2}{\partial x} - \frac{\partial u_1}{\partial y} \right) dS = \oint_C (u_1\, dx + u_2\, dy).$$

We will first prove Green's theorem for a rectangle with sides parallel to the axes. Let the rectangle be defined by $a \le x \le b$ and $c \le y \le d$, as pictured here. The area integral is given by

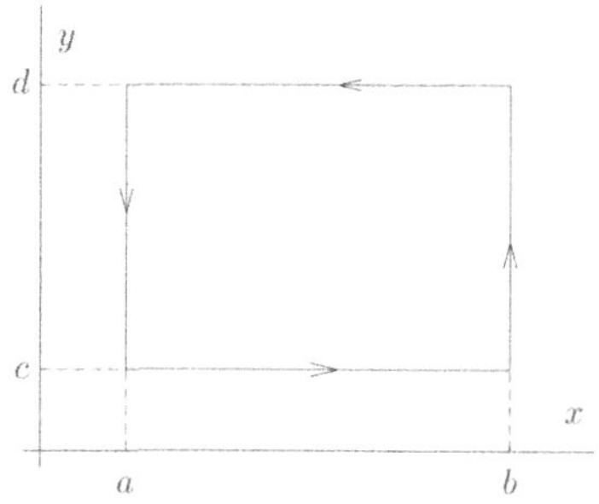

$$\int_S \left(\frac{\partial u_2}{\partial x} - \frac{\partial u_1}{\partial y} \right) dS$$
$$= \int_c^d \int_a^b \frac{\partial u_2}{\partial x} dx\, dy - \int_a^b \int_c^d \frac{\partial u_1}{\partial y} dy\, dx.$$

The inner integrals can be done using the fundamental theorem of calculus, and we obtain

$$\int_S \left(\frac{\partial u_2}{\partial x} - \frac{\partial u_1}{\partial y} \right) dS = \int_c^d \left[u_2(b,y) - u_2(a,y) \right] dy + \int_a^b \left[u_1(x,c) - u_1(x,d) \right] dx$$
$$= \oint_C (u_1\, dx + u_2\, dy).$$

Note that the line integral is done so that the bounded area is always to the left, which means counterclockwise.

Now, given any closed smooth curve in the x-y plane enclosing an area, we can subdivide the area by a very fine two-dimensional rectangular grid and apply the above result to each rectangle in the grid. All the area integrals over the rectangles add, whereas all the line integrals over the internal sides of the rectangles cancel. The only remaining line integrals are on the perimeter and approximate the given bounding curve. The result is Green's theorem for any area S in the plane bounded by a curve C.

Problems for Lecture 50

1. Test Green's theorem using $\boldsymbol{u} = -y\boldsymbol{i} + x\boldsymbol{j}$ for a square of side L lying in the first quadrant with vertex at the origin.

2. Test Green's theorem using $\boldsymbol{u} = -y\boldsymbol{i} + x\boldsymbol{j}$ for a circle of radius R centered at the origin.

Lecture 51 | Stokes' theorem

Green's theorem for a differentiable two-dimensional vector field,

$$\boldsymbol{u} = u_1(x,y)\,\boldsymbol{i} + u_2(x,y)\,\boldsymbol{j},$$

and a smooth curve C in the x-y plane surrounding an area S is given by

$$\int_S \left(\frac{\partial u_2}{\partial x} - \frac{\partial u_1}{\partial y} \right) dS = \oint_C (u_1\,dx + u_2\,dy).$$

Green's theorem can be extended to three dimensions. With

$$\boldsymbol{u} = u_1(x,y,z)\,\boldsymbol{i} + u_2(x,y,z)\,\boldsymbol{j} + u_3(x,y,z)\,\boldsymbol{k},$$

we see that

$$\frac{\partial u_2}{\partial x} - \frac{\partial u_1}{\partial y} = (\boldsymbol{\nabla} \times \boldsymbol{u}) \cdot \boldsymbol{k}.$$

And with

$$d\boldsymbol{S} = \boldsymbol{k}\,dS, \qquad u_1\,dx + u_2\,dy = \boldsymbol{u} \cdot d\boldsymbol{r},$$

Green's theorem can be rewritten in the form

$$\int_S (\boldsymbol{\nabla} \times \boldsymbol{u}) \cdot d\boldsymbol{S} = \oint_C \boldsymbol{u} \cdot d\boldsymbol{r}.$$

This restatement of Green's theorem, if interpreted as an equation in three dimensions, is called Stokes' theorem. Here, S is a general three-dimensional surface bounded by a closed spatial curve C. A simple example would be a hemisphere located anywhere in space bounded by a circle.

The orientation of the closed curve and the normal vector to the surface should follow the right-hand rule. If your fingers of your right hand point in the direction of the line integral, your thumb should point in the direction of the normal vector to the surface.

Problems for Lecture 51

1. From Stokes' theorem, determine the form of Green's theorem for a curve lying in the

a) y-z plane;

b) z-x plane.

2. Test Stokes' theorem using $\boldsymbol{u} = -y\boldsymbol{i} + x\boldsymbol{j}$ for a hemisphere of radius R with $z > 0$ bounded by a circle of radius R lying in the x-y plane with center at the origin.

3. Consider the two-dimensional velocity field of a fluid given by

$$\boldsymbol{u}(x, y) = \frac{\Gamma}{2\pi}\left(\frac{-y\boldsymbol{i} + x\boldsymbol{j}}{x^2 + y^2}\right).$$

a) Using cylindrical coordinates, show that

$$\boldsymbol{u}(\rho, \phi, z) = \frac{\Gamma\hat{\boldsymbol{\phi}}}{2\pi\rho}.$$

b) The vorticity field of the fluid is defined as $\boldsymbol{\omega} = \boldsymbol{\nabla} \times \boldsymbol{u}$. Using cylindrical coordinates, show that $\boldsymbol{\omega} = 0$ provided $\rho \neq 0$.

c) Using Stokes' theorem, show that the surface integral of the vorticity field over an area in the x-y plane containing the origin is equal to Γ, and therefore that the vorticity is given by $\boldsymbol{\omega} = \Gamma\delta(x)\delta(y)\hat{\boldsymbol{k}}$, where $\delta(x)$ and $\delta(y)$ are one-dimensional Dirac delta functions. This is the definition of a point vortex of strength Γ.

Practice Quiz | Stokes' theorem

1. Let $\boldsymbol{u} = -y\boldsymbol{i} + x\boldsymbol{j}$. Compute $\oint_C \boldsymbol{u} \cdot d\boldsymbol{r}$ for the quarter circle of radius R as illustrated. Here, it is simpler to apply Stokes' theorem to compute an area integral. The answer is

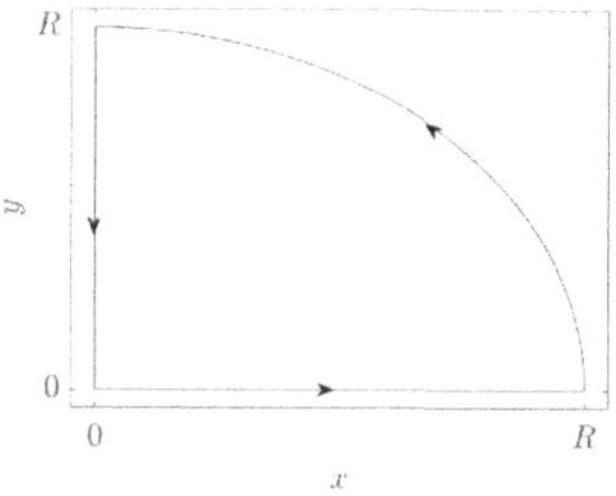

a) 0

b) $\frac{1}{2}\pi R^2$

c) πR^2

d) $2\pi R^2$

2. Let $\boldsymbol{u} = \dfrac{-y}{x^2 + y^2}\boldsymbol{i} + \dfrac{x}{x^2 + y^2}\boldsymbol{j}$. Compute the value of $\int_S (\boldsymbol{\nabla} \times \boldsymbol{u}) \cdot d\boldsymbol{S}$ over a circle of radius R centered at the origin in the x-y plane with normal vector $\boldsymbol{k}$. Here, because $\boldsymbol{u}$ is singular at $\boldsymbol{r} = 0$, it is necessary to apply Stokes' theorem and compute a line integral. The answer is

a) 0

b) π

c) 2π

d) 4π

3. Let $\boldsymbol{u} = -x^2y\boldsymbol{i} + xy^2\boldsymbol{j}$. Compute $\oint_C \boldsymbol{u} \cdot d\boldsymbol{r}$ for a unit square in the first quadrant with vertex at the origin. Here, it is simpler to compute an area integral. The answer is

a) 0

b) $\dfrac{1}{3}$

c) $\dfrac{2}{3}$

d) 1

Lecture 52 Meaning of the divergence and the curl

With $\boldsymbol{u}$ a differentiable vector field defined inside and on a smooth closed surface S enclosing a volume V, the divergence theorem states

$$\int_V \boldsymbol{\nabla} \cdot \boldsymbol{u} \, dV = \oint_S \boldsymbol{u} \cdot d\boldsymbol{S}.$$

We can limit this expression by shrinking the integration volume down to a point to obtain a coordinate-free representation of the divergence as

$$\boldsymbol{\nabla} \cdot \boldsymbol{u} = \lim_{V \to 0} \frac{1}{V} \oint_S \boldsymbol{u} \cdot d\boldsymbol{S}.$$

Picture V as the volume of a small sphere with surface S and $\boldsymbol{u}$ as the velocity field of some fluid of constant density. Then if the flow of fluid into the sphere is equal to the flow of fluid out of the sphere, the surface integral will be zero and $\boldsymbol{\nabla} \cdot \boldsymbol{u} = 0$. However, if more fluid flows out of the sphere than in, then $\boldsymbol{\nabla} \cdot \boldsymbol{u} > 0$ and if more fluid flows in than out, $\boldsymbol{\nabla} \cdot \boldsymbol{u} < 0$. Positive divergence indicates a source of fluid and negative divergence indicates a sink of fluid.

Now consider a surface S bounded by a curve C on which a differentiable vector field $\mathbf{u}$ is defined. Stokes' theorem states that

$$\int_S (\boldsymbol{\nabla} \times \boldsymbol{u}) \cdot d\boldsymbol{S} = \oint_C \boldsymbol{u} \cdot d\boldsymbol{r}.$$

We can limit this expression by shrinking the integration surface down to a point. With $\boldsymbol{n}$ a unit normal vector to the surface, with direction given by the right-hand rule, we obtain

$$(\boldsymbol{\nabla} \times \boldsymbol{u}) \cdot \boldsymbol{n} = \lim_{S \to 0} \frac{1}{S} \oint_C \boldsymbol{u} \cdot d\boldsymbol{r}.$$

Picture S as the area of a small disk bounded by a circle C and again picture $\boldsymbol{u}$ as the velocity field of a fluid. The line integral of $\boldsymbol{u} \cdot d\boldsymbol{r}$ around the circle C is called the flow's circulation and measures the swirl of the fluid around the center of the circle. The vector field $\boldsymbol{\omega} = \boldsymbol{\nabla} \times \boldsymbol{u}$ is called the vorticity of the fluid. The vorticity is most decidedly nonzero in a whirling (say, turbulent) fluid, composed of eddies of all different sizes.

Problems for Lecture 52

1. The incompressible Navier-Stokes equation governing fluid flow is given by

$$\frac{\partial \boldsymbol{u}}{\partial t} + (\boldsymbol{u} \cdot \boldsymbol{\nabla})\boldsymbol{u} = -\frac{1}{\rho}\boldsymbol{\nabla} p + \nu \nabla^2 \boldsymbol{u},$$

with $\boldsymbol{\nabla} \cdot \boldsymbol{u} = 0$. Here, $\boldsymbol{u}$, p, ρ and ν are the fluid's velocity, pressure, density and kinematic viscosity, respectively.

a) By taking the divergence of both sides of the Navier-Stokes equation, derive the following equation for the pressure in terms of the velocity field:

$$\nabla^2 p = -\rho \frac{\partial u_i}{\partial x_j}\frac{\partial u_j}{\partial x_i}.$$

b) By taking the curl of both sides of the Navier-Stokes equation, and defining the vorticity as $\boldsymbol{\omega} = \boldsymbol{\nabla} \times \boldsymbol{u}$, derive the vorticity equation

$$\frac{\partial \boldsymbol{\omega}}{\partial t} + (\boldsymbol{u} \cdot \boldsymbol{\nabla})\boldsymbol{\omega} = (\boldsymbol{\omega} \cdot \boldsymbol{\nabla})\boldsymbol{u} + \nu \nabla^2 \boldsymbol{\omega}.$$

You can use all the vector identities presented in these lecture notes, but you will need to prove that

$$\boldsymbol{u} \times (\boldsymbol{\nabla} \times \boldsymbol{u}) = \frac{1}{2}\boldsymbol{\nabla}(\boldsymbol{u} \cdot \boldsymbol{u}) - (\boldsymbol{u} \cdot \boldsymbol{\nabla})\boldsymbol{u}.$$

Lecture 53 Maxwell's equations

Maxwell's equations in SI units and in integral form are given by

$$\oint_S \boldsymbol{E} \cdot d\boldsymbol{S} = \frac{q_{\text{enc}}}{\varepsilon_0}, \qquad \text{(Gauss's law for electric fields)}$$

$$\oint_S \boldsymbol{B} \cdot d\boldsymbol{S} = 0, \qquad \text{(Gauss's law for magnetic fields)}$$

$$\oint_C \boldsymbol{E} \cdot d\boldsymbol{r} = -\frac{d}{dt}\int_S \boldsymbol{B} \cdot d\boldsymbol{S}, \qquad \text{(Faraday's law)}$$

$$\oint_C \boldsymbol{B} \cdot d\boldsymbol{r} = \mu_0 \left(I_{\text{enc}} + \varepsilon_0 \frac{d}{dt}\int_S \boldsymbol{E} \cdot d\boldsymbol{S} \right), \qquad \text{(Ampère-Maxwell law)}$$

where $\boldsymbol{E}$ and $\boldsymbol{B}$ are the electric and magnetic fields, q_{enc} is the charge enclosed by the bounding surface S and I_{enc} is the current through the bounding surface. The dimensional constants ε_0 and μ_0 are called the permittivity and permeability of free space.

The transformation from integral to differential form is a straightforward application of both the divergence and Stokes' theorem. The charge q_{enc} in the volume V and the current I_{enc} through the surface S are related to the charge density ρ and the current density $\boldsymbol{J}$ by

$$q_{\text{enc}} = \int_V \rho \, dV, \qquad I_{\text{enc}} = \int_S \boldsymbol{J} \cdot d\boldsymbol{S}.$$

We apply the divergence theorem to the surface integrals and Stokes' theorem to the line integrals, replace q_{enc} and I_{enc} by integrals over ρ and J, and combine the results to obtain

$$\int_V \left(\boldsymbol{\nabla} \cdot \boldsymbol{E} - \frac{\rho}{\varepsilon_0} \right) dV = 0, \qquad \int_S \left(\boldsymbol{\nabla} \times \boldsymbol{E} + \frac{\partial \boldsymbol{B}}{\partial t} \right) \cdot d\boldsymbol{S} = 0,$$

$$\int_V (\boldsymbol{\nabla} \cdot \boldsymbol{B}) \, dV = 0, \qquad \int_S \left(\boldsymbol{\nabla} \times \boldsymbol{B} - \mu_0 \left(\boldsymbol{J} + \varepsilon_0 \frac{\partial \boldsymbol{E}}{\partial t} \right) \right) \cdot d\boldsymbol{S} = 0.$$

Since the integration volumes and surfaces are of arbitrary size and shape, the integrands must vanish and we obtain the aesthetically appealing differential forms for Maxwell's equations given by

$$\boldsymbol{\nabla} \cdot \boldsymbol{E} = \frac{\rho}{\varepsilon_0}, \qquad \boldsymbol{\nabla} \times \boldsymbol{E} = -\frac{\partial \boldsymbol{B}}{\partial t},$$

$$\boldsymbol{\nabla} \cdot \boldsymbol{B} = 0, \qquad \boldsymbol{\nabla} \times \boldsymbol{B} = \mu_0 \left(\boldsymbol{J} + \varepsilon_0 \frac{\partial \boldsymbol{E}}{\partial t} \right).$$

Problems for Lecture 53

1. Using Gauss's law for the electric field given by

$$\oint_S \boldsymbol{E} \cdot d\boldsymbol{S} = \frac{q_{\text{enc}}}{\varepsilon_0},$$

determine the electric field of a point charge q at the origin. Assume the electric field is spherically symmetric.

2. Using Ampère's law,

$$\oint_C \boldsymbol{B} \cdot d\boldsymbol{r} = \mu_0 I_{\text{enc}},$$

(a special case of the Ampère-Maxwell law for static fields), determine the magnetic field of a current carrying infinite wire placed on the z-axis. Assume the magnetic field has cylindrical symmetry.

Appendices

Appendix A Matrix addition and multiplication

Two-by-two matrices A and B, with two rows and two columns, can be written as

$$\mathrm{A} = \begin{pmatrix} a_{11} & a_{12} \\ a_{21} & a_{22} \end{pmatrix}, \qquad \mathrm{B} = \begin{pmatrix} b_{11} & b_{12} \\ b_{21} & b_{22} \end{pmatrix}.$$

The first row of matrix A has elements a_{11} and a_{12}; the second row has elements a_{21} and a_{22}. The first column has elements a_{11} and a_{21}; the second column has elements a_{12} and a_{22}. Matrices can be multiplied by scalars and added. This is done element-by-element as follows:

$$k\mathrm{A} = \begin{pmatrix} ka_{11} & ka_{12} \\ ka_{21} & ka_{22} \end{pmatrix}, \qquad \mathrm{A} + \mathrm{B} = \begin{pmatrix} a_{11} + b_{11} & a_{12} + b_{12} \\ a_{21} + b_{21} & a_{22} + b_{22} \end{pmatrix}.$$

Matrices can also be multiplied. Matrix multiplication does not commute, and two matrices can be multiplied only if the number of columns of the matrix on the left equals the number of rows of the matrix on the right. One multiplies matrices by going across the rows of the first matrix and down the columns of the second matrix. The two-by-two example is given by

$$\begin{pmatrix} a_{11} & a_{12} \\ a_{21} & a_{22} \end{pmatrix} \begin{pmatrix} b_{11} & b_{12} \\ b_{21} & b_{22} \end{pmatrix} = \begin{pmatrix} a_{11}b_{11} + a_{12}b_{21} & a_{11}b_{12} + a_{12}b_{22} \\ a_{21}b_{11} + a_{22}b_{21} & a_{21}b_{12} + a_{22}b_{22} \end{pmatrix}.$$

Making use of the definition of matrix multiplication, a system of linear equations can be written in matrix form. For instance, a general system with two equations and two unknowns is given by

$$a_{11}x_1 + a_{12}x_2 = b_1, \qquad a_{21}x_1 + a_{22}x_2 = b_2;$$

and the matrix form of this equation is given by

$$\begin{pmatrix} a_{11} & a_{12} \\ a_{21} & a_{22} \end{pmatrix} \begin{pmatrix} x_1 \\ x_2 \end{pmatrix} = \begin{pmatrix} b_1 \\ b_2 \end{pmatrix}.$$

In short, this matrix equation is commonly written as

$$\mathrm{Ax} = \mathrm{b}.$$

Appendix B Matrix determinants and inverses

We denote the inverse of an n-by-n matrix A as A^{-1}, where

$$AA^{-1} = A^{-1}A = I,$$

and where I is the n-by-n identity matrix satisfying $IA = AI = A$. In particular, if A is an invertible matrix, then the unique solution to the matrix equation $Ax = b$ is given by $x = A^{-1}b$.

It can be shown that a matrix A is invertible if and only if its determinant is not zero. Here, we only need two-by-two and three-by-three determinants. The two-by-two determinant, using the vertical bar notation, is given by

$$\begin{vmatrix} a_{11} & a_{12} \\ a_{21} & a_{22} \end{vmatrix} = a_{11}a_{22} - a_{12}a_{21};$$

that is, multiply the diagonal elements and subtract the product of the off-diagonal elements.

The three-by-three determinant is given in terms of two-by-two determinants as

$$\begin{vmatrix} a_{11} & a_{12} & a_{13} \\ a_{21} & a_{22} & a_{23} \\ a_{31} & a_{32} & a_{33} \end{vmatrix} = a_{11}\begin{vmatrix} a_{22} & a_{23} \\ a_{32} & a_{33} \end{vmatrix} - a_{12}\begin{vmatrix} a_{21} & a_{23} \\ a_{31} & a_{33} \end{vmatrix} + a_{13}\begin{vmatrix} a_{21} & a_{22} \\ a_{31} & a_{32} \end{vmatrix}.$$

The rule here is to go across the first row of the matrix, multiplying each element in the row by the determinant of the matrix obtained by crossing out that element's row and column, and adding the results with alternating signs.

We will need to invert two-by-two and three-by-three matrices, but this will mainly be simple because our matrices will be orthogonal. The rows (or columns) of an orthogonal matrix, considered as components of a vector, are orthonormal. For example, the following two matrices are orthogonal matrices:

$$\begin{pmatrix} \cos\theta & \sin\theta \\ -\sin\theta & \cos\theta \end{pmatrix}, \qquad \begin{pmatrix} \sin\theta\cos\phi & \sin\theta\sin\phi & \cos\theta \\ \cos\theta\cos\phi & \cos\theta\sin\phi & -\sin\theta \\ -\sin\phi & \cos\phi & 0 \end{pmatrix}.$$

For the first matrix, the row vectors $\hat{\boldsymbol{r}} = \cos\theta\boldsymbol{i} + \sin\theta\boldsymbol{j}$ and $\hat{\boldsymbol{\theta}} = -\sin\theta\boldsymbol{i} + \cos\theta\boldsymbol{j}$ have unit length and are orthogonal, and the same can be said for the rows of the second matrix.

The inverse of an orthogonal matrix is simply given by its transpose, obtained by interchanging the matrices rows and columns. For example,

$$\begin{pmatrix} \cos\theta & \sin\theta \\ -\sin\theta & \cos\theta \end{pmatrix}^{-1} = \begin{pmatrix} \cos\theta & -\sin\theta \\ \sin\theta & \cos\theta \end{pmatrix}.$$

For more general two-by-two matrices, the inverse can be found from

$$\begin{pmatrix} a & b \\ c & d \end{pmatrix}^{-1} = \frac{1}{ad - bc} \begin{pmatrix} d & -b \\ -c & a \end{pmatrix},$$

i.e., switch the diagonal elements, negate the off-diagonal elements, and divide by the determinant.

Appendix C | Problem and practice quiz solutions

Solutions to the Problems for Lecture 1

1. We show the associative law graphically:

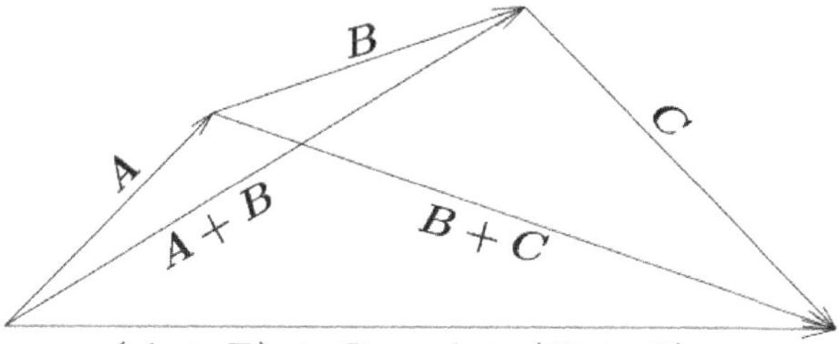

2.

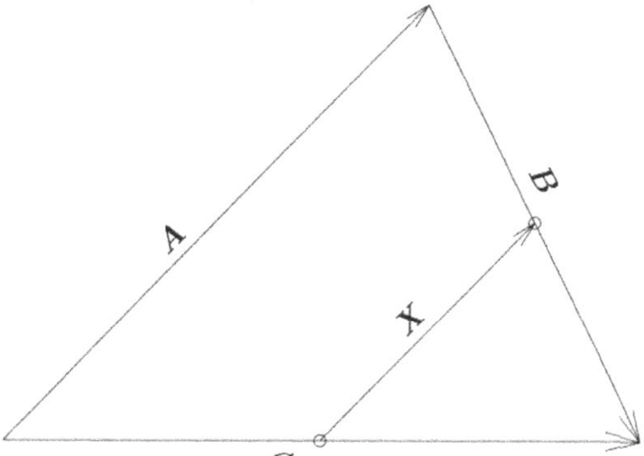

Draw a triangle with sides composed of the vectors $\boldsymbol{A}$, $\boldsymbol{B}$, and $\boldsymbol{C}$, with $\boldsymbol{C} = \boldsymbol{A} + \boldsymbol{B}$. Then draw the vector $\boldsymbol{X}$ pointing from the midpoint of $\boldsymbol{C}$ to the midpoint of $\boldsymbol{B}$.

From the figure, we see that

$$\frac{1}{2}\boldsymbol{C} + \boldsymbol{X} = \boldsymbol{A} + \frac{1}{2}\boldsymbol{B}.$$

Using $\boldsymbol{C} = \boldsymbol{A} + \boldsymbol{B}$, this equation becomes

$$\frac{1}{2}(\boldsymbol{A} + \boldsymbol{B}) + \boldsymbol{X} = \boldsymbol{A} + \frac{1}{2}\boldsymbol{B},$$

and solving for $\boldsymbol{X}$ yields $\boldsymbol{X} = \frac{1}{2}\boldsymbol{A}$. Therefore $\boldsymbol{X}$ is parallel to $\boldsymbol{A}$ and one-half its length.

Solutions to the Problems for Lecture 2

1. The unit vector that points from m_1 to m_2 is given by

$$\frac{\boldsymbol{r}_2 - \boldsymbol{r}_1}{|\boldsymbol{r}_2 - \boldsymbol{r}_1|} = \frac{(x_2 - x_1)\boldsymbol{i} + (y_2 - y_1)\boldsymbol{j} + (z_2 - z_1)\boldsymbol{k}}{\sqrt{(x_2 - x_1)^2 + (y_2 - y_1)^2 + (z_2 - z_1)^2}}.$$

2. The force acting on m_1 with position vector $\boldsymbol{r}_1$ due to the mass m_2 with position vector $\boldsymbol{r}_2$ is written as

$$\boldsymbol{F} = Gm_1m_2\frac{\boldsymbol{r}_2 - \boldsymbol{r}_1}{|\boldsymbol{r}_2 - \boldsymbol{r}_1|^3} = Gm_1m_2\frac{(x_2 - x_1)\boldsymbol{i} + (y_2 - y_1)\boldsymbol{j} + (z_2 - z_1)\boldsymbol{k}}{[(x_2 - x_1)^2 + (y_2 - y_1)^2 + (z_2 - z_1)^2]^{3/2}}.$$

Solutions to the Problems for Lecture 3

1.

a) $\boldsymbol{A} \cdot \boldsymbol{B} = A_1B_1 + A_2B_2 + A_3B_3 = B_1A_1 + B_2A_2 + B_3A_3 = \boldsymbol{B} \cdot \boldsymbol{A}$;

b) $\boldsymbol{A} \cdot (\boldsymbol{B} + \boldsymbol{C}) = A_1(B_1 + C_1) + A_2(B_2 + C_2) + A_3(B_3 + C_3) = A_1B_1 + A_1C_1 + A_2B_2 + A_2C_2 + A_3B_3 + A_3C_3 = (A_1B_1 + A_2B_2 + A_3B_3) + (A_1C_1 + A_2C_2 + A_3C_3) = \boldsymbol{A} \cdot \boldsymbol{B} + \boldsymbol{A} \cdot \boldsymbol{C}$;

c) $\boldsymbol{A} \cdot (k\boldsymbol{B}) = A_1(kB_1) + A_2(kB_2) + A_3(kB_3) = (kA_1)B_1 + (kA_2)B_2 + (kA_3)B_3 = k(A_1B_1) + k(A_2B_2) + k(A_3B_3) = (k\boldsymbol{A}) \cdot \boldsymbol{B} = k(\boldsymbol{A} \cdot \boldsymbol{B})$

2. The dot product of a unit vector with itself is one, and the dot product of a unit vector with one perpendicular to itself is zero. That is,

$$\boldsymbol{i} \cdot \boldsymbol{i} = \boldsymbol{j} \cdot \boldsymbol{j} = \boldsymbol{k} \cdot \boldsymbol{k} = 1; \qquad \boldsymbol{i} \cdot \boldsymbol{j} = \boldsymbol{i} \cdot \boldsymbol{k} = \boldsymbol{j} \cdot \boldsymbol{k} = 0; \qquad \boldsymbol{j} \cdot \boldsymbol{i} = \boldsymbol{k} \cdot \boldsymbol{i} = \boldsymbol{k} \cdot \boldsymbol{j} = 0.$$

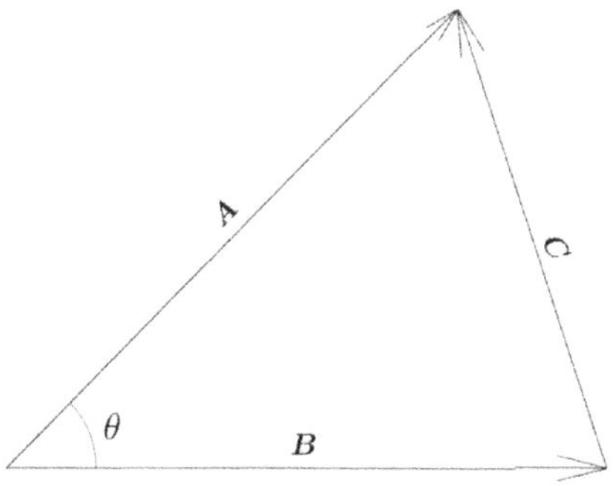

3. Consider the triangle composed of three vectors pictured above.

With $\boldsymbol{C} = \boldsymbol{A} - \boldsymbol{B}$, we have

$$\begin{aligned}|\boldsymbol{C}|^2 = \boldsymbol{C}\cdot\boldsymbol{C} = (\boldsymbol{A}-\boldsymbol{B})\cdot(\boldsymbol{A}-\boldsymbol{B}) &= \boldsymbol{A}\cdot\boldsymbol{A} + \boldsymbol{B}\cdot\boldsymbol{B} - 2\boldsymbol{A}\cdot\boldsymbol{B} \\ &= |\boldsymbol{A}|^2 + |\boldsymbol{B}|^2 - 2|\boldsymbol{A}||\boldsymbol{B}|\cos\theta,\end{aligned}$$

where θ is the angle between vectors $\boldsymbol{A}$ and $\boldsymbol{B}$. In the usual notation, if A, B and C are the lengths of the sides of a triangle, and θ is the angle opposite side C, then

$$C^2 = A^2 + B^2 - 2AB\cos\theta.$$

Solutions to the Problems for Lecture 4

1.

a)

$$\boldsymbol{A}\times\boldsymbol{B} = \begin{vmatrix} \boldsymbol{i} & \boldsymbol{j} & \boldsymbol{k} \\ A_1 & A_2 & A_3 \\ B_1 & B_2 & B_3 \end{vmatrix} = -\begin{vmatrix} \boldsymbol{i} & \boldsymbol{j} & \boldsymbol{k} \\ B_1 & B_2 & B_3 \\ A_1 & A_2 & A_3 \end{vmatrix} = -\boldsymbol{B}\times\boldsymbol{A}.$$

b)

$$\begin{aligned}\boldsymbol{A}\times(\boldsymbol{B}+\boldsymbol{C}) &= \begin{vmatrix} \boldsymbol{i} & \boldsymbol{j} & \boldsymbol{k} \\ A_1 & A_2 & A_3 \\ B_1+C_1 & B_2+C_2 & B_3+C_3 \end{vmatrix} \\ &= \begin{vmatrix} \boldsymbol{i} & \boldsymbol{j} & \boldsymbol{k} \\ A_1 & A_2 & A_3 \\ B_1 & B_2 & B_3 \end{vmatrix} + \begin{vmatrix} \boldsymbol{i} & \boldsymbol{j} & \boldsymbol{k} \\ A_1 & A_2 & A_3 \\ C_1 & C_2 & C_3 \end{vmatrix} = \boldsymbol{A}\times\boldsymbol{B} + \boldsymbol{A}\times\boldsymbol{C}.\end{aligned}$$

c)

$$\boldsymbol{A} \times (k\boldsymbol{B}) = \begin{vmatrix} \boldsymbol{i} & \boldsymbol{j} & \boldsymbol{k} \\ A_1 & A_2 & A_3 \\ kB_1 & kB_2 & kB_3 \end{vmatrix} = k \begin{vmatrix} \boldsymbol{i} & \boldsymbol{j} & \boldsymbol{k} \\ A_1 & A_2 & A_3 \\ B_1 & B_2 & B_3 \end{vmatrix} = \begin{vmatrix} \boldsymbol{i} & \boldsymbol{j} & \boldsymbol{k} \\ kA_1 & kA_2 & kA_3 \\ B_1 & B_2 & B_3 \end{vmatrix}$$
$$= k(\boldsymbol{A} \times \boldsymbol{B}) = (k\boldsymbol{A}) \times \boldsymbol{B}.$$

2. The cross product of a unit vector with itself is equal to the zero vector, the cross product of a unit vector with another (keeping the order cyclical in $\boldsymbol{i}, \boldsymbol{j}, \boldsymbol{k}$) is equal to the third unit vector, and reversing the order of multiplication changes the sign. That is,

$$\boldsymbol{i} \times \boldsymbol{i} = 0, \quad \boldsymbol{j} \times \boldsymbol{j} = 0, \quad \boldsymbol{k} \times \boldsymbol{k} = 0;$$
$$\boldsymbol{i} \times \boldsymbol{j} = \boldsymbol{k}, \quad \boldsymbol{j} \times \boldsymbol{k} = \boldsymbol{i}, \quad \boldsymbol{k} \times \boldsymbol{i} = \boldsymbol{j};$$
$$\boldsymbol{k} \times \boldsymbol{j} = -\boldsymbol{i}, \quad \boldsymbol{j} \times \boldsymbol{i} = -\boldsymbol{k}, \quad \boldsymbol{i} \times \boldsymbol{k} = -\boldsymbol{j}.$$

3. One such example is

$$\boldsymbol{i} \times (\boldsymbol{i} \times \boldsymbol{k}) = -\boldsymbol{i} \times \boldsymbol{j} = -\boldsymbol{k},$$
$$(\boldsymbol{i} \times \boldsymbol{i}) \times \boldsymbol{k} = 0 \times \boldsymbol{k} = 0.$$

Solutions to the Practice quiz: Vectors

1. c. As an example, $\boldsymbol{i} \times (\boldsymbol{i} \times \boldsymbol{j}) \neq (\boldsymbol{i} \times \boldsymbol{i}) \times \boldsymbol{j}$.

2. b.

$$(\boldsymbol{A} \times \boldsymbol{B}) \cdot \boldsymbol{j} = \begin{vmatrix} \boldsymbol{i} & \boldsymbol{j} & \boldsymbol{k} \\ a_1 & a_2 & a_3 \\ b_1 & b_2 & b_3 \end{vmatrix} \cdot \boldsymbol{j} = a_3 b_1 - a_1 b_3.$$

3. d.

$$\boldsymbol{i} \times (\boldsymbol{j} \times \boldsymbol{k}) = \boldsymbol{i} \times \boldsymbol{i} = 0, \quad (\boldsymbol{i} \times \boldsymbol{j}) \times \boldsymbol{k} = \boldsymbol{k} \times \boldsymbol{k} = 0,$$
$$(\boldsymbol{i} \times \boldsymbol{i}) \times \boldsymbol{j} = 0 \times \boldsymbol{j} = 0, \quad \boldsymbol{i} \times (\boldsymbol{i} \times \boldsymbol{j}) = \boldsymbol{i} \times \boldsymbol{k} = -\boldsymbol{j}.$$

Solutions to the Problems for Lecture 5

1. We first compute the displacement vector between $(1,1,1)$ and $(2,3,2)$:

$$\boldsymbol{u} = (2-1)\boldsymbol{i} + (3-1)\boldsymbol{j} + (2-1)\boldsymbol{k} = \boldsymbol{i} + 2\boldsymbol{j} + \boldsymbol{k}.$$

Choosing a point on the line to be $(1,1,1)$, the parametric equation for the line is given by

$$\boldsymbol{r} = \boldsymbol{r}_0 + \boldsymbol{u}t = (\boldsymbol{i}+\boldsymbol{j}+\boldsymbol{k}) + (\boldsymbol{i}+2\boldsymbol{j}+\boldsymbol{k})t = (1+t)\boldsymbol{i} + (1+2t)\boldsymbol{j} + (1+t)\boldsymbol{k}.$$

The line crosses the $x = 0$ and $z = 0$ planes when $t = -1$ at the intersection point $(0,-1,0)$, and crosses the $y = 0$ plane when $t = -1/2$ at the intersection point $(1/2, 0, 1/2)$.

Solutions to the Problems for Lecture 6

1. We find two vectors parallel to the plane defined by the three points, $(-1,-1,-1)$, $(1,1,1)$, and $(1,-1,0)$:

$$\boldsymbol{s}_1 = (1+1)\boldsymbol{i} + (1+1)\boldsymbol{j} + (1+1)\boldsymbol{k} = 2\boldsymbol{i} + 2\boldsymbol{j} + 2\boldsymbol{k},$$
$$\boldsymbol{s}_2 = (1-1)\boldsymbol{i} + (-1-1)\boldsymbol{j} + (0-1)\boldsymbol{k} = -2\boldsymbol{j} - \boldsymbol{k}.$$

We can divide $\boldsymbol{s}_1$ by 2 to construct a normal vector from

$$\boldsymbol{N} = \frac{1}{2}\boldsymbol{s}_1 \times \boldsymbol{s}_2 = \begin{vmatrix} \boldsymbol{i} & \boldsymbol{j} & \boldsymbol{k} \\ 1 & 1 & 1 \\ 0 & -2 & -1 \end{vmatrix} = \boldsymbol{i} + \boldsymbol{j} - 2\boldsymbol{k}.$$

The equation for the plane can be found from $\boldsymbol{N} \cdot (\boldsymbol{r} - \boldsymbol{r}_2) = 0$, or $\boldsymbol{N} \cdot \boldsymbol{r} = \boldsymbol{N} \cdot \boldsymbol{r}_2$, or

$$(\boldsymbol{i}+\boldsymbol{j}-2\boldsymbol{k}) \cdot (x\boldsymbol{i}+y\boldsymbol{j}+z\boldsymbol{k}) = (\boldsymbol{i}+\boldsymbol{j}-2\boldsymbol{k}) \cdot (\boldsymbol{i}+\boldsymbol{j}+\boldsymbol{k}), \quad \text{or} \quad x+y-2z = 0.$$

The intersection of this plane with the $z = 0$ plane forms the line given by $y = -x$.

Solutions to the Practice quiz: Analytic geometry

1. d. Write the parametric equation as $\boldsymbol{r} = \boldsymbol{r}_0 + \boldsymbol{u}t$. Using the point $(0,1,1)$, we take $\boldsymbol{r}_0 = \boldsymbol{j} + \boldsymbol{k}$ and from both points $(0,1,1)$ and $(1,0,-1)$, we have $\boldsymbol{u} = (1-0)\boldsymbol{i} + (0-1)\boldsymbol{j} + (-1-1)\boldsymbol{k} = \boldsymbol{i} - \boldsymbol{j} - 2\boldsymbol{k}$. Therefore $\boldsymbol{r} = \boldsymbol{j} + \boldsymbol{k} + (\boldsymbol{i} - \boldsymbol{j} - 2\boldsymbol{k})t = t\boldsymbol{i} + (1-t)\boldsymbol{j} + (1-2t)\boldsymbol{k}$.

2. a. The line is parameterized as $\boldsymbol{r} = t\boldsymbol{i} + (1-t)\boldsymbol{j} + (1-2t)\boldsymbol{k}$. The intersection with the $z = 0$ plane occurs when $t = 1/2$ so that $\boldsymbol{r} = \frac{1}{2}\boldsymbol{i} + \frac{1}{2}\boldsymbol{j}$. The intersection point is therefore $(\frac{1}{2}, \frac{1}{2}, 0)$.

3. d. We first find the parametric equation for the plane. From the points $(1,1,1)$, $(1,1,2)$ and $(2,1,1)$, we construct the two displacement vectors

$$\boldsymbol{s}_1 = (1-1)\boldsymbol{i} + (1-1)\boldsymbol{j} + (2-1)\boldsymbol{k} = \boldsymbol{k}$$
$$\boldsymbol{s}_2 = (2-1)\boldsymbol{i} + (1-1)\boldsymbol{j} + (1-2)\boldsymbol{k} = \boldsymbol{i} - \boldsymbol{k}.$$

The normal vector to the plane can be found from

$$\boldsymbol{N} = \boldsymbol{s}_1 \times \boldsymbol{s}_2 = \boldsymbol{k} \times (\boldsymbol{i} - \boldsymbol{k}) = \boldsymbol{k} \times \boldsymbol{i} - \boldsymbol{k} \times \boldsymbol{k} = \boldsymbol{j}.$$

Therefore, the parametric equation for the plane, given by $\boldsymbol{N} \cdot (\boldsymbol{r} - \boldsymbol{r}_1) = 0$, is determined to be

$$\boldsymbol{j} \cdot ((x-1)\boldsymbol{i} + (y-1)\boldsymbol{j} + (z-1)\boldsymbol{k}) = 0,$$

or $y = 1$. This plane is parallel to the x-z plane and when $z = 0$ is simply the line $y = 1$ for all values of x. Note now that we could have guessed this result because all three points defining the plane are located at $y = 1$.

Solutions to the Problems for Lecture 7

1.

a) If ijk is a cyclic permutation of $(1,2,3)$, then $\epsilon_{ijk} = \epsilon_{jki} = \epsilon_{kij} = 1$. If ijk is an anticyclic permutation of $(1,2,3)$, then $\epsilon_{ijk} = \epsilon_{jki} = \epsilon_{kij} = -1$. And if any two indices are equal, then $\epsilon_{ijk} = \epsilon_{jki} = \epsilon_{kij} = 0$. The use is that we can cyclically permute the indices of the Levi-Civita symbol without changing its value.

b) If ijk is a cyclic permutation of $(1,2,3)$, then $\epsilon_{ijk} = 1$ and $\epsilon_{jik} = \epsilon_{kji} = \epsilon_{ikj} = -1$. If ijk is an anticyclic permutation of $(1,2,3)$, then $\epsilon_{ijk} = -1$ and $\epsilon_{jik} = \epsilon_{kji} = \epsilon_{ikj} = 1$. And if any two indices are equal, then $\epsilon_{ijk} = \epsilon_{jik} = \epsilon_{ikj} = 0$. The use is that we can swap any two indices of the Levi-Civita symbol if we change its sign.

2. Notice that in the expression $\epsilon_{ijk}A_jB_k$, the indices j and k are repeated (and therefore summed over) but the index i is not. Taking $i = 1, 2,$ or 3, we have

$$\begin{aligned}
\epsilon_{1jk}A_jB_k &= \epsilon_{123}A_2B_3 + \epsilon_{132}A_3B_2 = A_2B_3 - A_3B_2 = [\boldsymbol{A}\times\boldsymbol{B}]_1,\\
\epsilon_{2jk}A_jB_k &= \epsilon_{231}A_3B_1 + \epsilon_{213}A_1B_3 = A_3B_1 - A_1B_3 = [\boldsymbol{A}\times\boldsymbol{B}]_2,\\
\epsilon_{3jk}A_jB_k &= \epsilon_{312}A_1B_2 + \epsilon_{321}A_2B_1 = A_1B_2 - A_2B_1 = [\boldsymbol{A}\times\boldsymbol{B}]_3.
\end{aligned}$$

3.

a) Now, $\delta_{ij}A_j = \delta_{i1}A_1 + \delta_{i2}A_2 + \delta_{i3}A_3$. The only nonzero term has the index of A equal to i, therefore $\delta_{ij}A_j = A_i$.

b) Now, $\delta_{ik}\delta_{kj} = \delta_{i1}\delta_{1j} + \delta_{i2}\delta_{2j} + \delta_{i3}\delta_{3j}$. If $i \neq j$, then every term in the sum is zero. If $i = j$, then only one term is nonzero and equal to one. Therefore, $\delta_{ik}\delta_{kj} = \delta_{ij}$. This result could also be viewed as an application of Part (a).

4. We make use of the identities $\delta_{ii} = 3$ and $\delta_{ik}\delta_{jk} = \delta_{ij}$. For the Kronecker delta, the order of the indices doesn't matter. We also use

$$\epsilon_{ijk}\epsilon_{lmn} = \delta_{il}(\delta_{jm}\delta_{kn} - \delta_{jn}\delta_{km}) - \delta_{im}(\delta_{jl}\delta_{kn} - \delta_{jn}\delta_{kl}) + \delta_{in}(\delta_{jl}\delta_{km} - \delta_{jm}\delta_{kl}).$$

a)

$$\begin{aligned}
\epsilon_{ijk}\epsilon_{imn} &= \delta_{ii}(\delta_{jm}\delta_{kn} - \delta_{jn}\delta_{km}) - \delta_{im}(\delta_{ji}\delta_{kn} - \delta_{jn}\delta_{ki}) + \delta_{in}(\delta_{ji}\delta_{km} - \delta_{jm}\delta_{ki})\\
&= 3(\delta_{jm}\delta_{kn} - \delta_{jn}\delta_{km}) - (\delta_{jm}\delta_{kn} - \delta_{jn}\delta_{km}) + (\delta_{jn}\delta_{km} - \delta_{jm}\delta_{kn})\\
&= \delta_{jm}\delta_{kn} - \delta_{jn}\delta_{km}.
\end{aligned}$$

b) We use the result of a) and find

$$\epsilon_{ijk}\epsilon_{ijn} = \delta_{jj}\delta_{kn} - \delta_{jn}\delta_{kj} = 3\delta_{kn} - \delta_{kn} = 2\delta_{kn}.$$

Solutions to the Problems for Lecture 8

1. We sometimes need parentheses because the vector product is not associative and expressions can be evaluated in more than one way with different results. We can write without any ambiguity,

(scalar triple product)	$\boldsymbol{A}\cdot\boldsymbol{B}\times\boldsymbol{C} = \boldsymbol{B}\cdot\boldsymbol{C}\times\boldsymbol{A} = \boldsymbol{C}\cdot\boldsymbol{A}\times\boldsymbol{B}$,
(vector triple product)	$\boldsymbol{A}\times(\boldsymbol{B}\times\boldsymbol{C}) = \boldsymbol{A}\cdot\boldsymbol{C}\boldsymbol{B} - \boldsymbol{A}\cdot\boldsymbol{B}\boldsymbol{C}$,
(scalar quadruple product)	$\boldsymbol{A}\times\boldsymbol{B}\cdot\boldsymbol{C}\times\boldsymbol{D} = \boldsymbol{A}\cdot\boldsymbol{C}\boldsymbol{B}\cdot\boldsymbol{D} - \boldsymbol{A}\cdot\boldsymbol{D}\boldsymbol{B}\cdot\boldsymbol{C}$.

For the vector quadruple product, we can write without ambiguity

$$(\boldsymbol{A} \times \boldsymbol{B}) \times (\boldsymbol{C} \times \boldsymbol{D}) = (\boldsymbol{A} \times \boldsymbol{B}) \cdot \boldsymbol{D}\boldsymbol{C} - (\boldsymbol{A} \times \boldsymbol{B}) \cdot \boldsymbol{C}\boldsymbol{D},$$

or

$$(\boldsymbol{A} \times \boldsymbol{B}) \times (\boldsymbol{C} \times \boldsymbol{D}) = (\boldsymbol{A} \times \boldsymbol{B} \cdot \boldsymbol{D})\boldsymbol{C} - (\boldsymbol{A} \times \boldsymbol{B} \cdot \boldsymbol{C})\boldsymbol{D}.$$

The absence of any parentheses on the right-hand side would result in an ambiguity, however, since in general,

$$((\boldsymbol{A} \times \boldsymbol{B}) \cdot \boldsymbol{D})\boldsymbol{C} \neq \boldsymbol{A} \times ((\boldsymbol{B} \cdot \boldsymbol{D})\boldsymbol{C}).$$

The vector on the left-hand side of this expression is parallel to $\boldsymbol{C}$ while the vector on the right-hand side is perpendicular to $\boldsymbol{C}$. The two expressions can therefore be only equal if $\boldsymbol{C}$ is the zero vector.

Solutions to the Problems for Lecture 9

1. Consider $\boldsymbol{A} \cdot (\boldsymbol{B} \times \boldsymbol{C})$. If $\boldsymbol{A} = \boldsymbol{B}$, then $\boldsymbol{A} \cdot (\boldsymbol{B} \times \boldsymbol{C}) = \boldsymbol{A} \cdot (\boldsymbol{A} \times \boldsymbol{C}) = 0$ since $\boldsymbol{A} \times \boldsymbol{C}$ is orthogonal to $\boldsymbol{A}$. A similar results holds for $\boldsymbol{A} = \boldsymbol{C}$. If $\boldsymbol{B} = \boldsymbol{C}$, then $\boldsymbol{A} \cdot (\boldsymbol{B} \times \boldsymbol{C}) = \boldsymbol{A} \cdot (\boldsymbol{B} \times \boldsymbol{B}) = 0$ since $\boldsymbol{B} \times \boldsymbol{B} = 0$.

2. We use the fact that the value of the scalar triple product is unchanged under a cyclic permutation of the three vectors, and that the dot product is commutative. We have

$$\boldsymbol{A} \cdot \boldsymbol{B} \times \boldsymbol{C} = \boldsymbol{C} \cdot \boldsymbol{A} \times \boldsymbol{B} = \boldsymbol{A} \times \boldsymbol{B} \cdot \boldsymbol{C}.$$

3. Consider the scalar triple product

$$\boldsymbol{e}_i \cdot (\boldsymbol{e}_j \times \boldsymbol{e}_k).$$

We have already proved that if any two indices are equal, then the scalar triple product is zero. Furthermore, the scalar triple product is unchanged under a cyclic permutation of the three vectors so that

$$\boldsymbol{e}_1 \cdot (\boldsymbol{e}_2 \times \boldsymbol{e}_3) = \boldsymbol{e}_2 \cdot (\boldsymbol{e}_3 \times \boldsymbol{e}_1) = \boldsymbol{e}_3 \cdot (\boldsymbol{e}_1 \times \boldsymbol{e}_2),$$

and

$$\boldsymbol{e}_1 \cdot (\boldsymbol{e}_2 \times \boldsymbol{e}_3) = \boldsymbol{i} \cdot (\boldsymbol{j} \times \boldsymbol{k}) = \boldsymbol{i} \cdot \boldsymbol{i} = 1.$$

Finally,

$$\boldsymbol{e}_3 \cdot (\boldsymbol{e}_2 \times \boldsymbol{e}_1) = \boldsymbol{e}_2 \cdot (\boldsymbol{e}_1 \times \boldsymbol{e}_3) = \boldsymbol{e}_1 \cdot (\boldsymbol{e}_3 \times \boldsymbol{e}_2),$$

and

$$e_3 \cdot (e_2 \times e_1) = \boldsymbol{k} \cdot (\boldsymbol{j} \times \boldsymbol{i}) = -\boldsymbol{k} \cdot \boldsymbol{k} = -1.$$

We have computed all possible cases, and have thus proved

$$e_i \cdot (e_j \times e_k) = \epsilon_{ijk}.$$

A slicker proof would denote the mth component of e_j by e_{jm} and use $e_{jm} = \delta_{jm}$, and so on for the other unit vectors and components. Then

$$e_i \cdot (e_j \times e_k) = e_{il}\epsilon_{lmn}e_{jm}e_{kn} = \epsilon_{lmn}\delta_{il}\delta_{jm}\delta_{kn} = \epsilon_{ijk}.$$

Solutions to the Problems for Lecture 10

1. We prove the Jacobi identity using the vector triple product and rearranging terms:

$$\begin{aligned}
&\boldsymbol{A} \times (\boldsymbol{B} \times \boldsymbol{C}) + \boldsymbol{B} \times (\boldsymbol{C} \times \boldsymbol{A}) + \boldsymbol{C} \times (\boldsymbol{A} \times \boldsymbol{B}) \\
&= [(\boldsymbol{A}\cdot\boldsymbol{C})\boldsymbol{B} - (\boldsymbol{A}\cdot\boldsymbol{B})\boldsymbol{C}] + [(\boldsymbol{B}\cdot\boldsymbol{A})\boldsymbol{C} - (\boldsymbol{B}\cdot\boldsymbol{C})\boldsymbol{A}] + [(\boldsymbol{C}\cdot\boldsymbol{B})\boldsymbol{A} - (\boldsymbol{C}\cdot\boldsymbol{A})\boldsymbol{B}] \\
&= [(\boldsymbol{A}\cdot\boldsymbol{C})\boldsymbol{B} - (\boldsymbol{C}\cdot\boldsymbol{A})\boldsymbol{B}] + [(\boldsymbol{B}\cdot\boldsymbol{A})\boldsymbol{C} - (\boldsymbol{A}\cdot\boldsymbol{B})\boldsymbol{C}] + [(\boldsymbol{C}\cdot\boldsymbol{B})\boldsymbol{A} - (\boldsymbol{B}\cdot\boldsymbol{C})\boldsymbol{A}] \\
&= 0.
\end{aligned}$$

2. We want to prove that the scalar quadruple product satisfies

$$(\boldsymbol{A} \times \boldsymbol{B}) \cdot (\boldsymbol{C} \times \boldsymbol{D}) = (\boldsymbol{A}\cdot\boldsymbol{C})(\boldsymbol{B}\cdot\boldsymbol{D}) - (\boldsymbol{A}\cdot\boldsymbol{D})(\boldsymbol{B}\cdot\boldsymbol{C}).$$

We have

	Justification
$(\boldsymbol{A} \times \boldsymbol{B}) \cdot (\boldsymbol{C} \times \boldsymbol{D})$	
$= [\boldsymbol{A} \times \boldsymbol{B}]_i[\boldsymbol{C} \times \boldsymbol{D}]_i$	$\boldsymbol{X} \cdot \boldsymbol{Y} = X_iY_i$
$= \epsilon_{ijk}A_jB_k\epsilon_{ilm}C_lD_m$	$[\boldsymbol{X} \times \boldsymbol{Y}]_i = \epsilon_{ijk}X_jY_k$
$= \epsilon_{ijk}\epsilon_{ilm}A_jB_kC_lD_m$	commutative law
$= (\delta_{jl}\delta_{km} - \delta_{jm}\delta_{kl})A_jB_kC_lD_m$	$\epsilon_{ijk}\epsilon_{ilm} = \delta_{jl}\delta_{km} - \delta_{jm}\delta_{kl}$
$= A_jC_jB_kD_k - A_jD_jB_kC_k$	$\delta_{jl}C_l = C_j$, $\delta_{km}D_m = D_k$, etc.
$= (\boldsymbol{A}\cdot\boldsymbol{C})(\boldsymbol{B}\cdot\boldsymbol{D}) - (\boldsymbol{A}\cdot\boldsymbol{D})(\boldsymbol{B}\cdot\boldsymbol{C})$.	$A_jC_j = \boldsymbol{A}\cdot\boldsymbol{C}$, $B_kD_k = \boldsymbol{B}\cdot\boldsymbol{D}$, etc.

3. We can prove Lagrange's identity using the scalar quadruple product identity $(\boldsymbol{A} \times \boldsymbol{B}) \cdot (\boldsymbol{C} \times \boldsymbol{D}) = (\boldsymbol{A} \cdot \boldsymbol{C})(\boldsymbol{B} \cdot \boldsymbol{D}) - (\boldsymbol{A} \cdot \boldsymbol{D})(\boldsymbol{B} \cdot \boldsymbol{C})$. We have

$$\begin{aligned}|\boldsymbol{A} \times \boldsymbol{B}|^2 = (\boldsymbol{A} \times \boldsymbol{B}) \cdot (\boldsymbol{A} \times \boldsymbol{B}) = (\boldsymbol{A} \cdot \boldsymbol{A})(\boldsymbol{B} \cdot \boldsymbol{B}) - (\boldsymbol{A} \cdot \boldsymbol{B})(\boldsymbol{B} \cdot \boldsymbol{A}) \\ = |\boldsymbol{A}|^2|\boldsymbol{B}|^2 - (\boldsymbol{A} \cdot \boldsymbol{B})^2.\end{aligned}$$

An alternative proof uses

$$\begin{aligned}|\boldsymbol{A} \times \boldsymbol{B}|^2 &= |\boldsymbol{A}|^2|\boldsymbol{B}|^2 \sin^2\theta = |\boldsymbol{A}|^2|\boldsymbol{B}|^2(1 - \cos^2\theta) \\ &= |\boldsymbol{A}|^2|\boldsymbol{B}|^2 - |\boldsymbol{A}|^2|\boldsymbol{B}|^2\cos^2\theta = |\boldsymbol{A}|^2|\boldsymbol{B}|^2 - (\boldsymbol{A} \cdot \boldsymbol{B})^2.\end{aligned}$$

4. We want to prove that the vector quadruple product satisfies

$$(\boldsymbol{A} \times \boldsymbol{B}) \times (\boldsymbol{C} \times \boldsymbol{D}) = ((\boldsymbol{A} \times \boldsymbol{B}) \cdot \boldsymbol{D})\boldsymbol{C} - ((\boldsymbol{A} \times \boldsymbol{B}) \cdot \boldsymbol{C})\boldsymbol{D}.$$

We will make use of the vector triple product identity given by

$$\boldsymbol{A} \times (\boldsymbol{B} \times \boldsymbol{C}) = (\boldsymbol{A} \cdot \boldsymbol{C})\boldsymbol{B} - (\boldsymbol{A} \cdot \boldsymbol{B})\boldsymbol{C}.$$

Let $\mathbf{X} = \boldsymbol{A} \times \boldsymbol{B}$. Then using the vector triple product identity, we have

$$\begin{aligned}(\boldsymbol{A} \times \boldsymbol{B}) \times (\boldsymbol{C} \times \boldsymbol{D}) &= \boldsymbol{X} \times (\boldsymbol{C} \times \boldsymbol{D}) \\ &= (\boldsymbol{X} \cdot \boldsymbol{D})\boldsymbol{C} - (\boldsymbol{X} \cdot \boldsymbol{C})\boldsymbol{D} \\ &= ((\boldsymbol{A} \times \boldsymbol{B}) \cdot \boldsymbol{D})\boldsymbol{C} - ((\boldsymbol{A} \times \boldsymbol{B}) \cdot \boldsymbol{C})\boldsymbol{D}.\end{aligned}$$

Solutions to the Practice quiz: Vector algebra

1. c. The relevant formula from the lecture is $\epsilon_{ijk}\epsilon_{ilm} = \delta_{jl}\delta_{km} - \delta_{jm}\delta_{kl}$. To directly apply this formula, we permute the indices of the Levi-Civita symbols without changing their cyclic order:

$$\epsilon_{ijk}\epsilon_{ljm} = \epsilon_{jki}\epsilon_{jml} = \delta_{km}\delta_{il} - \delta_{kl}\delta_{im}.$$

2. d. The other expressions can be shown to be false using $\boldsymbol{A} \times \boldsymbol{B} = -\boldsymbol{B} \times \boldsymbol{A}$ and, in general, $\boldsymbol{A} \times (\boldsymbol{B} \times \boldsymbol{C}) \neq (\boldsymbol{A} \times \boldsymbol{B}) \times \boldsymbol{C}$.

3. c. Use the facts that $\boldsymbol{A} \times \boldsymbol{B}$ is orthogonal to both $\boldsymbol{A}$ and $\boldsymbol{B}$, $\boldsymbol{A} \cdot \boldsymbol{B}$ is zero if $\boldsymbol{A}$ and $\boldsymbol{B}$ are orthogonal, and $\boldsymbol{A} \times \boldsymbol{B}$ is zero if $\boldsymbol{A}$ and $\boldsymbol{B}$ are parallel.

Solutions to the Problems for Lecture 11

1. *Scalar fields*: electrostatic potential, gravitational potential, temperature, humidity, concentration, density, pressure, wavefunction of quantum mechanics.
Vector fields: electric and magnetic fields, magnetic vector potential, velocity, force fields such as gravity.

Solutions to the Problems for Lecture 12

1. Using the chain rule,

$$\frac{\partial f}{\partial x} = \frac{-2nx}{(x^2+y^2+z^2)^{n+1}}, \quad \frac{\partial f}{\partial y} = \frac{-2ny}{(x^2+y^2+z^2)^{n+1}}, \quad \frac{\partial f}{\partial z} = \frac{-2nz}{(x^2+y^2+z^2)^{n+1}}.$$

2. Define

$$f(t+\epsilon, x+\delta) = g(\epsilon, \delta).$$

Then the first-order Taylor series expansion of g is given by

$$g(\epsilon, \delta) = g(0,0) + \epsilon g_t(0,0) + \delta g_x(0,0),$$

which in terms of f becomes

$$f(t+\epsilon, x+\delta) = f(t,x) + \epsilon f_t(t,x) + \delta f_x(t,x).$$

Applying this expansion to $f(t+\alpha\Delta t, x+\beta\Delta t f(t,x))$, we have to first-order in Δt,

$$f(t+\alpha\Delta t, x+\beta\Delta t f(t,x)) = f(t,x) + \alpha\Delta t f_t(t,x) + \beta\Delta t f(t,x) f_x(t,x).$$

Solutions to the Problems for Lecture 13

1. The formulas derived in the text are

$$\beta_0 = \frac{\sum x_i^2 \sum y_i - \sum x_i y_i \sum x_i}{n\sum x_i^2 - (\sum x_i)^2}, \quad \beta_1 = \frac{n\sum x_i y_i - (\sum x_i)(\sum y_i)}{n\sum x_i^2 - (\sum x_i)^2},$$

where the sum is from $i = 1$ to 3. Here, $x_1 = 1$, $x_2 = 2$, $x_3 = 3$, and $y_1 = 1$, $y_2 = 3$ and $y_3 = 2$. We have

$$\beta_0 = \frac{(14)(6) - (13)(6)}{(3)(14) - (6)^2} = 1, \qquad \beta_1 = \frac{(3)(13) - (6)(6)}{(3)(14) - (6)^2} = 1/2.$$

The best fit line is therefore $y = 1 + x/2$. The graph of the data and the line are shown below.

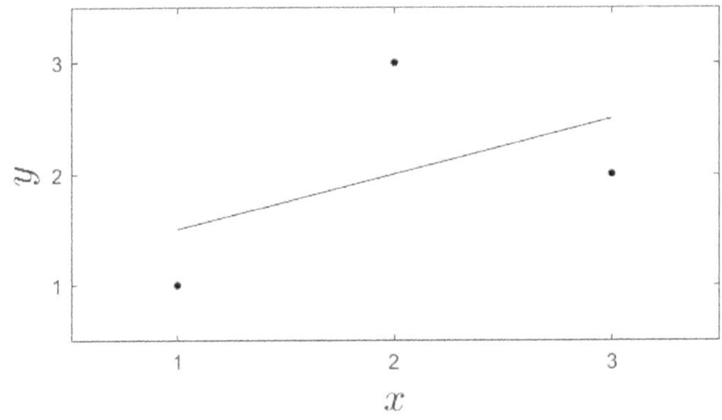

Solutions to the Problems for Lecture 14

1.

a) With $f(x, y) = e^{xy}$, $x = r\cos\theta$, and $y = r\sin\theta$, application of the chain rule results in

$$\begin{aligned}\frac{\partial f}{\partial r} &= \frac{\partial f}{\partial x}\frac{\partial x}{\partial r} + \frac{\partial f}{\partial y}\frac{\partial y}{\partial r}\\ &= ye^{xy}\cos\theta + xe^{xy}\sin\theta\\ &= r\sin\theta\cos\theta e^{r^2\cos\theta\sin\theta} + r\sin\theta\cos\theta e^{r^2\cos\theta\sin\theta}\\ &= 2r\sin\theta\cos\theta e^{r^2\cos\theta\sin\theta},\end{aligned}$$

and

$$\begin{aligned}\frac{\partial f}{\partial \theta} &= \frac{\partial f}{\partial x}\frac{\partial x}{\partial \theta} + \frac{\partial f}{\partial y}\frac{\partial y}{\partial \theta}\\ &= ye^{xy}(-r\sin\theta) + xe^{xy}(r\cos\theta)\\ &= -r^2\sin^2\theta e^{r^2\cos\theta\sin\theta} + r^2\cos^2\theta e^{r^2\cos\theta\sin\theta}\\ &= r^2(\cos^2\theta - \sin^2\theta)e^{r^2\cos\theta\sin\theta}.\end{aligned}$$

b) Substituting for x and y, we have $f = e^{r^2\cos\theta\sin\theta}$. Then

$$\begin{aligned}\frac{\partial f}{\partial r} &= 2r\cos\theta\sin\theta e^{r^2\cos\theta\sin\theta},\\ \frac{\partial f}{\partial \theta} &= r^2(\cos^2\theta - \sin^2\theta)e^{r^2\cos\theta\sin\theta}.\end{aligned}$$

Solutions to the Problems for Lecture 16

1. Suppose $ax + by + cz = 0$. We have the relations

$$x = \frac{-by - cz}{a}, \qquad y = \frac{-ax - cz}{b}, \qquad z = \frac{-ax - by}{c}.$$

The partial derivatives are

$$\frac{\partial x}{\partial y} = -\frac{b}{a}, \qquad \frac{\partial y}{\partial z} = -\frac{c}{b}, \qquad \frac{\partial z}{\partial x} = -\frac{a}{c};$$

and the triple product is

$$\frac{\partial x}{\partial y}\frac{\partial y}{\partial z}\frac{\partial z}{\partial x} = \left(-\frac{b}{a}\right)\left(-\frac{c}{b}\right)\left(-\frac{a}{c}\right) = -1.$$

2. Suppose $ax + by + cz + dt = 0$. We have the relations

$$x = \frac{-by - cz - dt}{a}, \qquad y = \frac{-ax - cz - dt}{b},$$
$$z = \frac{-ax - by - dt}{c}, \qquad t = \frac{-ax - by - cz}{d}.$$

The partial derivatives are

$$\frac{\partial x}{\partial y} = -\frac{b}{a}, \qquad \frac{\partial y}{\partial z} = -\frac{c}{b}, \qquad \frac{\partial z}{\partial t} = -\frac{d}{c}, \qquad \frac{\partial t}{\partial x} = -\frac{a}{d};$$

and the quadruple product is

$$\frac{\partial x}{\partial y}\frac{\partial y}{\partial z}\frac{\partial z}{\partial t}\frac{\partial t}{\partial x} = \left(-\frac{b}{a}\right)\left(-\frac{c}{b}\right)\left(-\frac{d}{c}\right)\left(-\frac{a}{d}\right) = 1.$$

Apparently an odd number of products yields -1 and an even number of products yields $+1$.

Solutions to the Practice quiz: Partial derivatives

1. d. The partial derivative with respect to x is given by

$$\frac{\partial f}{\partial x} = \frac{-x}{(x^2 + y^2 + z^2)^{3/2}};$$

and the mixed second partial derivative is then given by

$$\frac{\partial^2 f}{\partial x \partial y} = \frac{3xy}{(x^2+y^2+z^2)^{5/2}}$$

2. a. From the data points $(0,1)$, $(1,3)$, $(2,3)$ and $(3,4)$, we compute

$$\sum x_i = 6, \quad \sum x_i^2 = 14, \quad \sum y_i = 11, \quad \sum x_i y_i = 21.$$

Then using

$$\beta_0 = \frac{\sum x_i^2 \sum y_i - \sum x_i y_i \sum x_i}{n\sum x_i^2 - (\sum x_i)^2}, \qquad \beta_1 = \frac{n\sum x_i y_i - (\sum x_i)(\sum y_i)}{n\sum x_i^2 - (\sum x_i)^2},$$

we have

$$\beta_0 = \frac{(14)(11)-(21)(6)}{(4)(14)-(6)^2} = \frac{154-126}{56-36} = \frac{28}{20} = \frac{7}{5},$$
$$\beta_1 = \frac{(4)(21)-(6)(11)}{(4)(14)-(6)^2} = \frac{84-66}{56-36} = \frac{18}{20} = \frac{9}{10}.$$

The least-squares line is therefore given by $y = 7/5 + 9x/10$.

3. d. Let $f = f(x,y)$ with $x = r\cos\theta$ and $y = r\sin\theta$. Then application of the chain rule results in

$$\begin{aligned}\frac{\partial f}{\partial \theta} &= \frac{\partial f}{\partial x}\frac{\partial x}{\partial \theta} + \frac{\partial f}{\partial y}\frac{\partial y}{\partial \theta} \\ &= -r\sin\theta\frac{\partial f}{\partial x} + r\cos\theta\frac{\partial f}{\partial y} \\ &= -y\frac{\partial f}{\partial x} + x\frac{\partial f}{\partial y}.\end{aligned}$$

Solutions to the Problems for Lecture 17

1.

a) Let $\phi(x,y,z) = x^2+y^2+z^2$. The gradient is given by

$$\boldsymbol{\nabla}\phi = \boldsymbol{\nabla}(x^2+y^2+z^2) = 2x\boldsymbol{i} + 2y\boldsymbol{j} + 2z\boldsymbol{k}.$$

In terms of the position vector, we have

$$\boldsymbol{\nabla}(r^2) = 2\boldsymbol{r}.$$

b) Let $\phi(x,y,z) = \sqrt{x^2+y^2+z^2}$. The gradient is given by

$$\begin{aligned}\boldsymbol{\nabla}\phi &= \boldsymbol{\nabla}\sqrt{x^2+y^2+z^2} \\ &= \frac{x}{\sqrt{x^2+y^2+z^2}}\boldsymbol{i} + \frac{y}{\sqrt{x^2+y^2+z^2}}\boldsymbol{j} + \frac{z}{\sqrt{x^2+y^2+z^2}}\boldsymbol{k}.\end{aligned}$$

In terms of the position vector, we have

$$\boldsymbol{\nabla}(r) = \frac{\boldsymbol{r}}{r} = \hat{\boldsymbol{r}}.$$

c) Let $\phi(x,y,z) = \dfrac{1}{\sqrt{x^2+y^2+z^2}}$. The gradient is given by

$$\begin{aligned}\boldsymbol{\nabla}\phi &= \boldsymbol{\nabla}\left(\frac{1}{\sqrt{x^2+y^2+z^2}}\right) \\ &= -\frac{x}{(x^2+y^2+z^2)^{3/2}}\boldsymbol{i} - \frac{y}{(x^2+y^2+z^2)^{3/2}}\boldsymbol{j} - \frac{z}{(x^2+y^2+z^2)^{3/2}}\boldsymbol{k}.\end{aligned}$$

In terms of the position vector, we have

$$\boldsymbol{\nabla}\left(\frac{1}{r}\right) = -\frac{\boldsymbol{r}}{r^3}.$$

2. Following the pattern given by

$$\boldsymbol{\nabla}(r^2) = 2\boldsymbol{r}, \quad \boldsymbol{\nabla}(r) = \frac{\boldsymbol{r}}{r}, \quad \boldsymbol{\nabla}\left(\frac{1}{r}\right) = -\frac{\boldsymbol{r}}{r^3},$$

we guess that the general result is

$$\boldsymbol{\nabla}(r^n) = n\boldsymbol{r}r^{n-2}.$$

Solutions to the Problems for Lecture 18

1.

a) With $\boldsymbol{F} = xy\boldsymbol{i} + yz\boldsymbol{j} + zx\boldsymbol{k}$, we have

$$\begin{aligned}\boldsymbol{\nabla}\cdot\boldsymbol{F} &= \frac{\partial}{\partial x}(xy) + \frac{\partial}{\partial y}(yz) + \frac{\partial}{\partial z}(zx) \\ &= y + z + x = x + y + z.\end{aligned}$$

b) With $\boldsymbol{F} = yz\boldsymbol{i} + xz\boldsymbol{j} + xy\boldsymbol{k}$, we have

$$\nabla \cdot \boldsymbol{F} = \frac{\partial}{\partial x}(yz) + \frac{\partial}{\partial y}(xz) + \frac{\partial}{\partial z}(xy) = 0.$$

Solutions to the Problems for Lecture 19

1.

a) With $\boldsymbol{F} = xy\boldsymbol{i} + yz\boldsymbol{j} + zx\boldsymbol{k}$, we have

$$\nabla \times \boldsymbol{F} = \begin{vmatrix} \boldsymbol{i} & \boldsymbol{j} & \boldsymbol{k} \\ \partial/\partial x & \partial/\partial y & \partial/\partial z \\ xy & yz & zx \end{vmatrix} = -y\boldsymbol{i} - z\boldsymbol{j} - x\boldsymbol{k}.$$

b) With $\boldsymbol{F} = yz\boldsymbol{i} + xz\boldsymbol{j} + xy\boldsymbol{k}$, we have

$$\nabla \times \boldsymbol{F} = \begin{vmatrix} \boldsymbol{i} & \boldsymbol{j} & \boldsymbol{k} \\ \partial/\partial x & \partial/\partial y & \partial/\partial z \\ yz & xz & xy \end{vmatrix} = (x - x)\boldsymbol{i} + (y - y)\boldsymbol{j} + (z - z)\boldsymbol{k} = 0.$$

2. With $\boldsymbol{u} = u_1(x,y)\boldsymbol{i} + u_2(x,y)\boldsymbol{j}$, we have

$$\boldsymbol{\omega} = \nabla \times \boldsymbol{u} = \begin{vmatrix} \boldsymbol{i} & \boldsymbol{j} & \boldsymbol{k} \\ \partial/\partial x & \partial/\partial y & \partial/\partial z \\ u_1(x,y) & u_2(x,y) & 0 \end{vmatrix} = 0\boldsymbol{i} + 0\boldsymbol{j} + \left(\frac{\partial u_2}{\partial x} - \frac{\partial u_1}{\partial y}\right)\boldsymbol{k} = \omega_3\boldsymbol{k}.$$

Therefore, $\omega_3 = \partial u_2/\partial x - \partial u_1/\partial y$.

Solutions to the Problems for Lecture 20

1. We have

$$\nabla^2\left(\frac{1}{r}\right) = \frac{\partial^2}{\partial x^2}\left(\frac{1}{\sqrt{x^2+y^2+z^2}}\right) + \frac{\partial^2}{\partial y^2}\left(\frac{1}{\sqrt{x^2+y^2+z^2}}\right) + \frac{\partial^2}{\partial z^2}\left(\frac{1}{\sqrt{x^2+y^2+z^2}}\right).$$

We can compute the derivatives with respect to x and use symmetry to find the other two terms. We have

$$\frac{\partial}{\partial x}\left(\frac{1}{\sqrt{x^2+y^2+z^2}}\right) = \frac{-x}{(x^2+y^2+z^2)^{3/2}};$$

and

$$\begin{aligned}\frac{\partial}{\partial x}\left(\frac{-x}{(x^2+y^2+z^2)^{3/2}}\right) &= \frac{-(x^2+y^2+z^2)^{3/2}+3x^2(x^2+y^2+z^2)^{1/2}}{(x^2+y^2+z^2)^3} \\ &= -\frac{1}{(x^2+y^2+z^2)^{3/2}} + \frac{3x^2}{(x^2+y^2+z^2)^{5/2}}.\end{aligned}$$

It is easy to guess the derivative with respect to y and z, and we have

$$\begin{aligned}\nabla^2\left(\frac{1}{r}\right) &= -\frac{3}{(x^2+y^2+z^2)^{3/2}} + \frac{3(x^2+y^2+z^2)}{(x^2+y^2+z^2)^{5/2}} \\ &= -\frac{3}{(x^2+y^2+z^2)^{3/2}} + \frac{3}{(x^2+y^2+z^2)^{3/2}} = 0,\end{aligned}$$

a result only valid for $r \neq 0$.

Solutions to the Practice quiz: The del operator

1. d. We have

$$\begin{aligned}\boldsymbol{\nabla}\left(\frac{1}{r^2}\right) &= \boldsymbol{\nabla}\left(\frac{1}{x^2+y^2+z^2}\right) \\ &= \frac{-2x}{(x^2+y^2+z^2)^2}\boldsymbol{i} + \frac{-2y}{(x^2+y^2+z^2)^2}\boldsymbol{j} + \frac{-2z}{(x^2+y^2+z^2)^2}\boldsymbol{k} \\ &= -\frac{2\boldsymbol{r}}{r^4}.\end{aligned}$$

2. b. We use

$$\boldsymbol{\nabla}\cdot\boldsymbol{F} = \boldsymbol{\nabla}\cdot\left(\frac{x\boldsymbol{i}+y\boldsymbol{j}+z\boldsymbol{k}}{\sqrt{x^2+y^2+z^2}}\right).$$

Now,

$$\begin{aligned}\frac{\partial}{\partial x}\left(\frac{x}{\sqrt{x^2+y^2+z^2}}\right) &= \frac{\sqrt{x^2+y^2+z^2} - x^2(x^2+y^2+z^2)^{-1/2}}{x^2+y^2+z^2} \\ &= \frac{1}{r} - \frac{x^2}{r^3},\end{aligned}$$

and similarly for the partial derivatives with respect to y and z. Adding all three partial derivatives results in

$$\nabla \cdot \boldsymbol{F} = \frac{3}{r} - \frac{x^2+y^2+z^2}{r^3} = \frac{3}{r} - \frac{1}{r} = \frac{2}{r}.$$

3.

b. We have

$$\nabla \times \boldsymbol{r} = \begin{vmatrix} \boldsymbol{i} & \boldsymbol{j} & \boldsymbol{k} \\ \partial/\partial x & \partial/\partial y & \partial/\partial z \\ x & y & z \end{vmatrix} = 0.$$

Solutions to the Problems for Lecture 22

1.

a) To prove $\nabla \cdot (f\boldsymbol{u}) = \boldsymbol{u} \cdot \nabla f + f \nabla \cdot \boldsymbol{u}$, we compute

$$\begin{aligned} \nabla \cdot (f\boldsymbol{u}) &= \frac{\partial}{\partial x_i}(f u_i) && \text{(divergence in component notation)} \\ &= \frac{\partial f}{\partial x_i} u_i + f \frac{\partial u_i}{\partial x_i} && \text{(product rule for the derivative)} \\ &= \boldsymbol{u} \cdot \nabla f + f \nabla \cdot \boldsymbol{u}. && \text{(back to vector notation)} \end{aligned}$$

b) To prove $\nabla \times (\nabla \times \boldsymbol{u}) = \nabla(\nabla \cdot \boldsymbol{u}) - \nabla^2 \boldsymbol{u}$, we compute for the ith component:

$$\begin{aligned} [\nabla \times (\nabla \times \boldsymbol{u})]_i &= \epsilon_{ijk} \frac{\partial}{\partial x_j}\left(\epsilon_{klm} \frac{\partial u_m}{\partial x_l}\right) && \text{(curl in component notation)} \\ &= \epsilon_{ijk}\epsilon_{klm} \frac{\partial^2 u_m}{\partial x_j \partial x_l} && (\epsilon_{klm} \text{ doesn't depend on } x_j) \\ &= \epsilon_{kij}\epsilon_{klm} \frac{\partial^2 u_m}{\partial x_j \partial x_l} && (\epsilon_{ijk} = \epsilon_{kij}) \\ &= (\delta_{il}\delta_{jm} - \delta_{im}\delta_{jl}) \frac{\partial^2 u_m}{\partial x_j \partial x_l} && (\epsilon_{kij}\epsilon_{klm} = \delta_{il}\delta_{jm} - \delta_{im}\delta_{jl}) \\ &= \frac{\partial^2 u_j}{\partial x_j \partial x_i} - \frac{\partial^2 u_i}{\partial x_j \partial x_j} && \left(\delta_{il} \frac{\partial^2 u_m}{\partial x_j \partial x_l} = \frac{\partial^2 u_m}{\partial x_j \partial x_i}, \text{ etc.}\right) \\ &= [\nabla(\nabla \cdot \boldsymbol{u})]_i - [\nabla^2 \boldsymbol{u}]_i. && \text{(back to vector notation)} \end{aligned}$$

Therefore, $\nabla \times (\nabla \times \boldsymbol{u}) = \nabla(\nabla \cdot \boldsymbol{u}) - \nabla^2 \boldsymbol{u}$.

2.

a) With $d\boldsymbol{r}/dt = \boldsymbol{u}(t, \boldsymbol{r}(t))$, the component equations are given by

$$\frac{dx_1}{dt} = u_1(t; x_1, x_2, x_3), \quad \frac{dx_2}{dt} = u_2(t; x_1, x_2, x_3), \quad \frac{dx_3}{dt} = u_3(t; x_1, x_2, x_3).$$

b) Using the chain rule, we can compute the second derivative of x_1 as

$$\begin{aligned}\frac{d^2x_1}{dt^2} &= \frac{\partial u_1}{\partial t} + \frac{\partial u_1}{\partial x_1}\frac{dx_1}{dt} + \frac{\partial u_1}{\partial x_2}\frac{dx_2}{dt} + \frac{\partial u_1}{\partial x_3}\frac{dx_3}{dt} \\ &= \frac{\partial u_1}{\partial t} + u_1\frac{\partial u_1}{\partial x_1} + u_2\frac{\partial u_1}{\partial x_2} + u_3\frac{\partial u_1}{\partial x_3}.\end{aligned}$$

Similarly for x_2 and x_3, we have

$$\begin{aligned}\frac{d^2x_2}{dt^2} &= \frac{\partial u_2}{\partial t} + u_1\frac{\partial u_2}{\partial x_1} + u_2\frac{\partial u_2}{\partial x_2} + u_3\frac{\partial u_2}{\partial x_3}, \\ \frac{d^2x_3}{dt^2} &= \frac{\partial u_3}{\partial t} + u_1\frac{\partial u_3}{\partial x_1} + u_2\frac{\partial u_3}{\partial x_2} + u_3\frac{\partial u_3}{\partial x_3}.\end{aligned}$$

c) Using the operator

$$\boldsymbol{u} \cdot \boldsymbol{\nabla} = u_1\frac{\partial}{\partial x_1} + u_2\frac{\partial}{\partial x_2} + u_3\frac{\partial}{\partial x_3},$$

the three components can be combined into the vector expression

$$\frac{d^2\boldsymbol{r}}{dt^2} = \frac{\partial \boldsymbol{u}}{\partial t} + \boldsymbol{u} \cdot \boldsymbol{\nabla}\boldsymbol{u}.$$

This expression is called the material acceleration, and is found in the Navier-Stokes equation of fluid mechanics.

Solutions to the Problems for Lecture 23

1. Start with Maxwell's equations:

$$\boldsymbol{\nabla} \cdot \boldsymbol{E} = 0, \qquad \boldsymbol{\nabla} \cdot \boldsymbol{B} = 0, \qquad \boldsymbol{\nabla} \times \boldsymbol{E} = -\frac{\partial \boldsymbol{B}}{\partial t}, \qquad \boldsymbol{\nabla} \times \boldsymbol{B} = \mu_0\epsilon_0\frac{\partial \boldsymbol{E}}{\partial t}.$$

Take the curl of the fourth Maxwell's equation, and commute the time and space derivatives to obtain

$$\boldsymbol{\nabla} \times (\boldsymbol{\nabla} \times \boldsymbol{B}) = \mu_0\epsilon_0\frac{\partial}{\partial t}(\boldsymbol{\nabla} \times \boldsymbol{E}).$$

Apply the curl of the curl identity to obtain

$$\boldsymbol{\nabla}(\boldsymbol{\nabla}\cdot\boldsymbol{B}) - \nabla^2\boldsymbol{B} = \mu_0\epsilon_0\frac{\partial}{\partial t}(\boldsymbol{\nabla}\times\boldsymbol{E}).$$

Apply the second Maxwell's equation to the left-hand-side, and the third Maxwell's equation to the right-hand-side. Rearranging terms, we obtain the three-dimensional wave equation given by

$$\frac{\partial^2\boldsymbol{B}}{\partial t^2} = c^2\nabla^2\boldsymbol{B},$$

where $c = 1/\sqrt{\mu_0\epsilon_0}$.

Solutions to the Practice quiz: Vector calculus algebra

1. a. We make use of the vector identity

$$\boldsymbol{\nabla}(\boldsymbol{u}\cdot\boldsymbol{v}) = (\boldsymbol{u}\cdot\boldsymbol{\nabla})\boldsymbol{v} + (\boldsymbol{v}\cdot\boldsymbol{\nabla})\boldsymbol{u} + \boldsymbol{u}\times(\boldsymbol{\nabla}\times\boldsymbol{v}) + \boldsymbol{v}\times(\boldsymbol{\nabla}\times\boldsymbol{u}).$$

Setting $\boldsymbol{v} = \boldsymbol{u}$, we have

$$\boldsymbol{\nabla}(\boldsymbol{u}\cdot\boldsymbol{u}) = 2(\boldsymbol{u}\cdot\boldsymbol{\nabla})\boldsymbol{u} + 2\boldsymbol{u}\times(\boldsymbol{\nabla}\times\boldsymbol{u}).$$

Therefore,

$$\frac{1}{2}\boldsymbol{\nabla}(\boldsymbol{u}\cdot\boldsymbol{u}) = \boldsymbol{u}\times(\boldsymbol{\nabla}\times\boldsymbol{u}) + (\boldsymbol{u}\cdot\boldsymbol{\nabla})\boldsymbol{u}.$$

2. c. The curl of a gradient (a. and d.) and the divergence of a curl (b.) are zero. The divergence of a gradient (c) is the Laplacian and is not always zero.

3. b. With $\boldsymbol{E}(\boldsymbol{r},t) = \sin(z-ct)\boldsymbol{i}$, we have $\boldsymbol{\nabla}\cdot\boldsymbol{E} = 0$, and

$$\boldsymbol{\nabla}\times\boldsymbol{E} = \begin{vmatrix} \boldsymbol{i} & \boldsymbol{j} & \boldsymbol{k} \\ \partial/\partial x & \partial/\partial y & \partial/\partial z \\ \sin(z-ct) & 0 & 0 \end{vmatrix} = \cos(z-ct)\boldsymbol{j}.$$

Maxwell's equation $\boldsymbol{\nabla}\times\boldsymbol{E} = -\dfrac{\partial\boldsymbol{B}}{\partial t}$ then results in

$$\frac{\partial\boldsymbol{B}}{\partial t} = -\cos(z-ct)\boldsymbol{j},$$

which can be integrated (setting the constant to zero) to obtain

$$B = \frac{1}{c}\sin(z - ct)\boldsymbol{j}.$$

Solutions to the Problems for Lecture 24

1. In general, the mass of a solid with mass density $\rho = \rho(x, y, z)$ is given by

$$M = \iiint_V \rho(x, y, z)\, dx\, dy\, dz.$$

To determine the mass of the cube, we place our coordinate system so that one corner of the cube is at the origin and the adjacent corners are on the positive x, y and z axes. We assume that the density of the cube is only a function of z, with

$$\rho(z) = \rho_1 + \frac{z}{L}(\rho_2 - \rho_1).$$

The mass of the cube is then given by

$$\begin{aligned}
M &= \int_0^L \int_0^L \int_0^L \left[\rho_1 + \frac{z}{L}(\rho_2 - \rho_1)\right] dx\, dy\, dz \\
&= \int_0^L dx \int_0^L dy \int_0^L \left[\rho_1 + \frac{z}{L}(\rho_2 - \rho_1)\right] dz \\
&= L^2 \left[\rho_1 z + \frac{z^2}{2L}(\rho_2 - \rho_1)\right]_0^L \\
&= L^3 \left[\rho_1 + \frac{1}{2}(\rho_2 - \rho_1)\right] \\
&= \frac{1}{2} L^3 (\rho_1 + \rho_2).
\end{aligned}$$

Solutions to the Problems for Lecture 25

1. The figure illustrates the integral over x first and y second. With this order, the integral over the parallelogram is given by

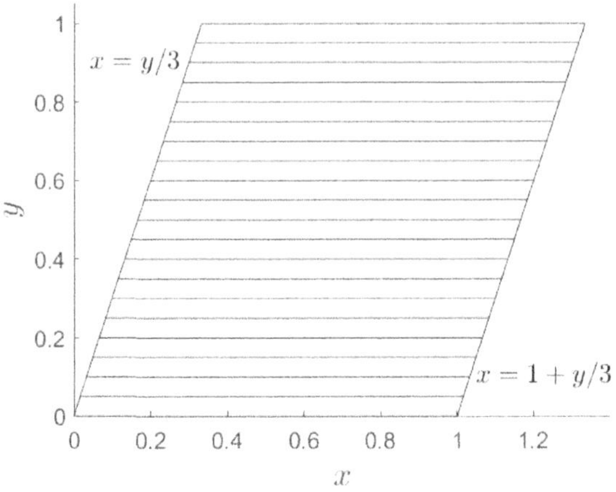

$$\begin{aligned}\int_0^1 \int_{y/3}^{1+y/3} x^2 y \, dx \, dy &= \int_0^1 \left. \frac{x^3 y}{3} \right|_{x=y/3}^{x=1+y/3} dy \\ &= \frac{1}{3} \int_0^1 y \left(\left(1 + \frac{1}{3}y\right)^3 - \left(\frac{1}{3}y\right)^3 \right) dy \\ &= \frac{1}{3} \int_0^1 y \left(1 + y + \frac{1}{3}y^2\right) dy \\ &= \frac{1}{3} \left(\frac{1}{2}y^2 + \frac{1}{3}y^3 + \frac{1}{12}y^4 \right) \bigg|_0^1 \\ &= \frac{1}{3} \left(\frac{1}{2} + \frac{1}{3} + \frac{1}{12} \right) = \frac{11}{36}.\end{aligned}$$

Solutions to the Practice quiz: Multidimensional integration

1. b. To find the volume, we integrate $z = xy$ over its base. We have

$$\int_0^1 \int_0^1 xy \, dx \, dy = \int_0^1 x \, dx \int_0^1 y \, dy = \left(\int_0^1 x \, dx \right)^2 = \left(\frac{1}{2} \right)^2 = \frac{1}{4}.$$

2. b. To determine the mass of the cube, we place our coordinate system so that one corner of the cube is at the origin and the adjacent corners are on the positive x, y and z axes. We assume that the density of the cube is only a function of z, with

$$\rho(z) = (1 + z) \, \text{g/cm}^3.$$

The mass of the cube in grams is then given by

$$M = \int_0^1 \int_0^1 \int_0^1 (1+z)\,dx\,dy\,dz$$
$$= \int_0^1 dx \int_0^1 dy \int_0^1 (1+z)\,dz = (z + \frac{1}{2}z^2)\Big|_0^1 = 1.5\,\text{g}.$$

3. d. We draw a picture of the triangle and illustrate the chosen direction of integration. Integrating first along x and then along y, the volume is given

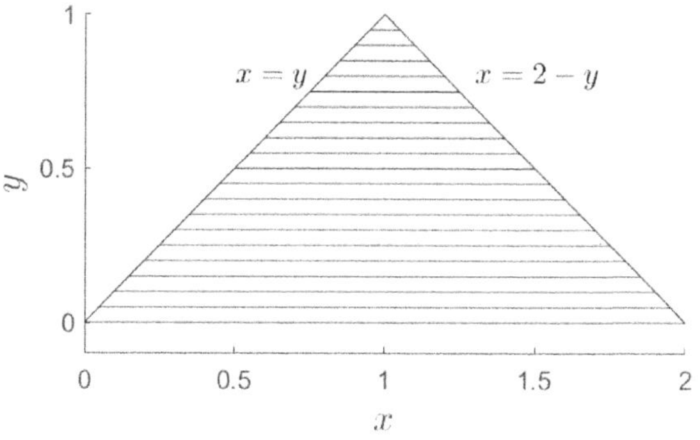

by

$$\int_0^1 \int_y^{2-y} xy\,dx\,dy = \int_0^1 \frac{1}{2}x^2\Big|_y^{2-y} y\,dy = \frac{1}{2}\int_0^1 y\left[(2-y)^2 - y^2\right] dy$$
$$= 2\int_0^1 (y - y^2)\,dy = 2\left(\frac{1}{2} - \frac{1}{3}\right) = \frac{1}{3}.$$

Solutions to the Problems for Lecture 26

1. The matrix form for the relationship between $\hat{r}$, $\hat{\theta}$ and i, j is given by

$$\begin{pmatrix} \hat{r} \\ \hat{\theta} \end{pmatrix} = \begin{pmatrix} \cos\theta & \sin\theta \\ -\sin\theta & \cos\theta \end{pmatrix} \begin{pmatrix} i \\ j \end{pmatrix}.$$

Inverting the two-by-two matrix, we have

$$\begin{pmatrix} \boldsymbol{i} \\ \boldsymbol{j} \end{pmatrix} = \begin{pmatrix} \cos\theta & -\sin\theta \\ \sin\theta & \cos\theta \end{pmatrix} \begin{pmatrix} \hat{\boldsymbol{r}} \\ \hat{\boldsymbol{\theta}} \end{pmatrix}.$$

Therefore,

$$\boldsymbol{i} = \cos\theta\hat{\boldsymbol{r}} - \sin\theta\hat{\boldsymbol{\theta}}, \qquad \boldsymbol{j} = \sin\theta\hat{\boldsymbol{r}} + \cos\theta\hat{\boldsymbol{\theta}}.$$

2.

a) With $x = r\cos\theta$ and $y = r\sin\theta$, and with $f = f(x(r,\theta), y(r,\theta))$, we have using the chain rule,

$$\begin{aligned} \frac{\partial f}{\partial r} &= \frac{\partial f}{\partial x}\frac{\partial x}{\partial r} + \frac{\partial f}{\partial y}\frac{\partial y}{\partial r} = \cos\theta\frac{\partial f}{\partial x} + \sin\theta\frac{\partial f}{\partial y}, \\ \frac{\partial f}{\partial \theta} &= \frac{\partial f}{\partial x}\frac{\partial x}{\partial \theta} + \frac{\partial f}{\partial y}\frac{\partial y}{\partial \theta} = -r\sin\theta\frac{\partial f}{\partial x} + r\cos\theta\frac{\partial f}{\partial y}. \end{aligned}$$

b) We can write the result of Part (a) in matrix form as

$$\begin{pmatrix} \partial f/\partial r \\ \partial f/\partial \theta \end{pmatrix} = \begin{pmatrix} \cos\theta & \sin\theta \\ -r\sin\theta & r\cos\theta \end{pmatrix} \begin{pmatrix} \partial f/\partial x \\ \partial f/\partial y \end{pmatrix}.$$

Inverting the two-by-two matrix results in

$$\begin{pmatrix} \partial f/\partial x \\ \partial f/\partial y \end{pmatrix} = \frac{1}{r} \begin{pmatrix} r\cos\theta & -\sin\theta \\ r\sin\theta & \cos\theta \end{pmatrix} \begin{pmatrix} \partial f/\partial r \\ \partial f/\partial \theta \end{pmatrix},$$

or

$$\frac{\partial f}{\partial x} = \cos\theta\frac{\partial f}{\partial r} - \frac{\sin\theta}{r}\frac{\partial f}{\partial \theta}, \qquad \frac{\partial f}{\partial y} = \sin\theta\frac{\partial f}{\partial r} + \frac{\cos\theta}{r}\frac{\partial f}{\partial \theta}.$$

The Cartesian partial derivatives in polar form are therefore

$$\frac{\partial}{\partial x} = \cos\theta\frac{\partial}{\partial r} - \frac{\sin\theta}{r}\frac{\partial}{\partial \theta}, \qquad \frac{\partial}{\partial y} = \sin\theta\frac{\partial}{\partial r} + \frac{\cos\theta}{r}\frac{\partial}{\partial \theta}.$$

3. We have

$$r\hat{\boldsymbol{r}} = r\cos\theta\,\boldsymbol{i} + r\sin\theta\,\boldsymbol{j} = x\boldsymbol{i} + y\boldsymbol{j},$$

and

$$r\hat{\boldsymbol{\theta}} = -r\sin\theta\,\boldsymbol{i} + r\cos\theta\,\boldsymbol{j} = -y\boldsymbol{i} + x\boldsymbol{j}.$$

Solutions to the Problems for Lecture 27

1. We have

$$\boldsymbol{u} = \frac{1}{r}\left(k_1\hat{\boldsymbol{r}} + k_2\hat{\boldsymbol{\theta}}\right) = u_r\hat{\boldsymbol{r}} + u_\theta\hat{\boldsymbol{\theta}},$$

so that

$$u_r = \frac{k_1}{r}, \qquad u_\theta = \frac{k_2}{r}.$$

The divergence is given by

$$\begin{aligned}\boldsymbol{\nabla}\cdot\boldsymbol{u} &= \frac{1}{r}\frac{\partial}{\partial r}(ru_r) + \frac{1}{r}\frac{\partial u_\theta}{\partial \theta}\\ &= \frac{1}{r}\frac{\partial}{\partial r}\left(r\left(\frac{k_1}{r}\right)\right) + \frac{1}{r}\frac{\partial}{\partial \theta}\left(\frac{k_2}{r}\right)\\ &= 0;\end{aligned}$$

and the curl is given by

$$\begin{aligned}\boldsymbol{\nabla}\times\boldsymbol{u} &= \boldsymbol{k}\left(\frac{1}{r}\frac{\partial}{\partial r}(ru_\theta) + \frac{1}{r}\frac{\partial u_r}{\partial \theta}\right)\\ &= \boldsymbol{k}\left(\frac{1}{r}\frac{\partial}{\partial r}\left(r\left(\frac{k_2}{r}\right)\right) + \frac{1}{r}\frac{\partial}{\partial \theta}\left(\frac{k_1}{r}\right)\right)\\ &= 0.\end{aligned}$$

It is important to emphasis that these results are only valid for $r \neq 0$. The vector field is singular when $r = 0$ and is not differentiable in the usual sense.

Solutions to the Problems for Lecture 28

1. With $u = u(r)$, we solve

$$\nabla^2 u = -\frac{G}{\nu\rho},$$

with boundary condition $u(R) = 0$. Writing the Laplacian in polar coordinates, we have

$$\frac{1}{r}\frac{d}{dr}\left(r\frac{du}{dr}\right) = -\frac{G}{\nu\rho}.$$

We multiply by r and integrate from 0 to r:

$$\int_0^r \frac{d}{dr}\left(r\frac{du}{dr}\right) dr = -\frac{G}{\nu\rho}\int_0^r r\,dr,$$

or

$$r\frac{du}{dr} = -\frac{Gr^2}{2\nu\rho}.$$

We now divide by r and integrate from r to R:

$$\int_r^R \frac{du}{dr}\,dr = -\frac{G}{2\nu\rho}\int_r^R r\,dr,$$

or

$$u(R) - u(r) = -\frac{G}{4\nu\rho}(R^2 - r^2).$$

Applying the no-slip boundary condition at $r = R$, we obtain

$$u(r) = \frac{GR^2}{4\nu\rho}\left(1 - \left(\frac{r}{R}\right)^2\right).$$

The maximum velocity occurs in the center of the pipe and is given by

$$u_{\mathrm{m}} = \frac{GR^2}{4\nu\rho},$$

so we can write the velocity field as

$$u(r) = u_{\mathrm{m}}\left(1 - \left(\frac{r}{R}\right)^2\right).$$

Solutions to the Problems for Lecture 29

1. We have

$$\boldsymbol{l} = \boldsymbol{r} \times \boldsymbol{p} = \boldsymbol{r} \times (m\dot{\boldsymbol{r}}) = mr\hat{\boldsymbol{r}} \times (\dot{r}\hat{\boldsymbol{r}} + r\dot{\theta}\hat{\boldsymbol{\theta}}) = mr^2\dot{\theta}(\hat{\boldsymbol{r}} \times \hat{\boldsymbol{\theta}}).$$

Now, $\hat{\boldsymbol{r}}$ and $\hat{\boldsymbol{\theta}}$ are perpendicular unit vectors so that $|\hat{\boldsymbol{r}} \times \hat{\boldsymbol{\theta}}| = 1$, and

$$|\boldsymbol{l}| = mr^2|\dot{\theta}||\hat{\boldsymbol{r}} \times \hat{\boldsymbol{\theta}}| = mr^2|\dot{\theta}|.$$

2. Using the Einstein summation convention, we have

$$\begin{aligned}\frac{d}{dt}\left(\frac{1}{2}m|\boldsymbol{v}|^2\right) &= \frac{1}{2}m\frac{d}{dt}(v_i v_i)\\ &= \frac{1}{2}m\left(\frac{dv_i}{dt}v_i + v_i\frac{dv_i}{dt}\right)\\ &= mv_i\frac{dv_i}{dt}\\ &= m\boldsymbol{v}\cdot\frac{d\boldsymbol{v}}{dt}.\end{aligned}$$

Solutions to the Problems for Lecture 31

1. The mass density of the disk is given by

$$\sigma(r) = \rho_0 + (\rho_1 - \rho_0)(r/R).$$

Integrating the mass density in polar coordinates to find the total mass of the disk, we have

$$\begin{aligned} M &= \int_0^{2\pi} \int_0^R [\rho_0 + (\rho_1 - \rho_0)(r/R)]\, r\, dr\, d\theta \\ &= 2\pi \left[\frac{\rho_0 r^2}{2} + \frac{(\rho_1 - \rho_0) r^3}{3R} \right]_{r=0}^{r=R} \\ &= \frac{1}{3}\pi R^2(\rho_0 + 2\rho_1). \end{aligned}$$

2. It is simplest to do this integral by transforming to polar coordinates. With $x^2 + y^2 = r^2$ and $dx\, dy = r\, dr\, d\theta$, we have

$$I^2 = \int_0^{2\pi} \int_0^\infty e^{-r^2} r\, dr\, d\theta = \int_0^{2\pi} d\theta \int_0^\infty r e^{-r^2}\, dr = 2\pi \int_0^\infty r e^{-r^2}\, dr.$$

Let $u = r^2$ and $du = 2r\, dr$. Then the integral transforms to

$$I^2 = \pi \int_0^\infty e^{-u}\, du = -\pi e^{-u} \Big|_0^\infty = \pi.$$

Therefore,

$$I = \int_{-\infty}^{\infty} e^{-x^2}\, dx = \sqrt{\pi}.$$

Solutions to the Practice quiz: Polar coordinates

1. d. Using $\hat{\boldsymbol{\theta}} = -\sin\theta \boldsymbol{i} + \cos\theta \boldsymbol{j}$, we have

$$r\hat{\boldsymbol{\theta}} = -r\sin\theta \boldsymbol{i} + r\cos\theta \boldsymbol{j} = -y\boldsymbol{i} + x\boldsymbol{j}.$$

2. b. With $\hat{\boldsymbol{r}} = \cos\theta \boldsymbol{i} + \sin\theta \boldsymbol{j}$ and $\hat{\boldsymbol{\theta}} = -\sin\theta \boldsymbol{i} + \cos\theta \boldsymbol{j}$, we have

$$\frac{d\hat{\boldsymbol{\theta}}}{d\theta} = -\cos\theta \boldsymbol{i} - \sin\theta \boldsymbol{j} = -\hat{\boldsymbol{r}}.$$

3. The mass density of the disk is given in polar coordinates by

$$\sigma = \sigma(r) = (10 - 9r)\,\mathrm{g/cm^2}.$$

The mass is found by integrating in polar coordinates using $dx\,dy = r\,dr\,d\theta$. Calculating in grams, we have

$$\begin{aligned} M &= \int_0^{2\pi}\int_0^1 (10-9r)r\,dr\,d\theta \\ &= \int_0^{2\pi} d\theta \int_0^1 (10-9r)r\,dr \\ &= 2\pi(5r^2 - 3r^3)\Big|_0^1 = 4\pi \approx 12.57\,\mathrm{g}. \end{aligned}$$

Solutions to the Problems for Lecture 32

1. We have

$$\begin{aligned} \boldsymbol{\nabla} &= \hat{\boldsymbol{x}}\frac{\partial}{\partial x} + \hat{\boldsymbol{y}}\frac{\partial}{\partial y} + \hat{\boldsymbol{z}}\frac{\partial}{\partial z} \\ &= (\cos\phi\hat{\boldsymbol{\rho}} - \sin\phi\hat{\boldsymbol{\phi}})\left(\cos\phi\frac{\partial}{\partial\rho} - \frac{\sin\phi}{\rho}\frac{\partial}{\partial\phi}\right) \\ &\quad + (\sin\phi\hat{\boldsymbol{\rho}} + \cos\phi\hat{\boldsymbol{\phi}})\left(\sin\phi\frac{\partial}{\partial\rho} + \frac{\cos\phi}{\rho}\frac{\partial}{\partial\phi}\right) + \hat{\boldsymbol{z}}\frac{\partial}{\partial z} \\ &= \hat{\boldsymbol{\rho}}\left(\cos^2\phi\frac{\partial}{\partial\rho} - \frac{\cos\phi\sin\phi}{\rho}\frac{\partial}{\partial\phi} + \sin^2\phi\frac{\partial}{\partial\rho} + \frac{\sin\phi\cos\phi}{\rho}\frac{\partial}{\partial\phi}\right) \\ &\quad + \hat{\boldsymbol{\phi}}\left(-\sin\phi\cos\phi\frac{\partial}{\partial\rho} + \frac{\sin^2\phi}{\rho}\frac{\partial}{\partial\phi} + \cos\phi\sin\phi\frac{\partial}{\partial\rho} + \frac{\cos^2\phi}{\rho}\frac{\partial}{\partial\phi}\right) + \hat{\boldsymbol{z}}\frac{\partial}{\partial z} \\ &= \hat{\boldsymbol{\rho}}\frac{\partial}{\partial\rho} + \hat{\boldsymbol{\phi}}\frac{1}{\rho}\frac{\partial}{\partial\phi} + \hat{\boldsymbol{z}}\frac{\partial}{\partial z}. \end{aligned}$$

2. The calculations are

a)

$$\boldsymbol{\nabla}\cdot\hat{\boldsymbol{\rho}} = \frac{1}{\rho}\frac{\partial}{\partial\rho}(\rho) = \frac{1}{\rho};$$

b)

$$
\begin{aligned}
\nabla \cdot \hat{\rho} &= \nabla \cdot (\cos\phi \boldsymbol{i} + \sin\phi \boldsymbol{j}) \\
&= \nabla \cdot \left(\frac{x}{\sqrt{x^2+y^2}} \boldsymbol{i} + \frac{y}{\sqrt{x^2+y^2}} \boldsymbol{j} \right) \\
&= \frac{\partial}{\partial x} \left(\frac{x}{\sqrt{x^2+y^2}} \right) + \frac{\partial}{\partial y} \left(\frac{y}{\sqrt{x^2+y^2}} \right) \\
&= \frac{\sqrt{x^2+y^2} - x^2(x^2+y^2)^{-1/2}}{x^2+y^2} + \frac{\sqrt{x^2+y^2} - y^2(x^2+y^2)^{-1/2}}{x^2+y^2} \\
&= \frac{2\sqrt{x^2+y^2} - \sqrt{x^2+y^2}}{x^2+y^2} \\
&= \frac{1}{\sqrt{x^2+y^2}} = \frac{1}{\rho}.
\end{aligned}
$$

3. $\nabla \times \hat{\rho} = 0$, $\nabla \cdot \hat{\phi} = 0$ and

$$
\nabla \times \hat{\phi} = \hat{\boldsymbol{z}} \frac{1}{\rho} \frac{\partial}{\partial \rho}(\rho) = \frac{1}{\rho} \hat{\boldsymbol{z}}.
$$

4. With ρ constant, the mass of the cone is given by $M = \rho V = \frac{1}{3}\pi\rho a^2 b$. By symmetry, we argue that $\boldsymbol{R} = Z\boldsymbol{k}$. Then

$$
Z = \frac{\rho}{M} \int_V z dV.
$$

We can perform the $dxdy$ integration at a fixed z by finding the area of a circular cross-section of the cone located at height z above the apex. The radius of this circular cross-section is given by $r = \frac{az}{b}$. Therefore, the volume integral reduces to a one-dimensional integral over z, and we have

$$
Z = \frac{\rho}{M} \int_0^b z\pi \left(\frac{az}{b} \right)^2 dz = \frac{\pi\rho a^2/b^2}{\frac{1}{3}\pi\rho a^2 b} \int_0^b z^3 dz = \frac{3}{b^3} \left(\frac{b^4}{4} \right) = \frac{3}{4} b.
$$

Solutions to the Problems for Lecture 33

1. The spherical coordinate unit vectors can be written in terms of the Cartesian unit vectors by

$$\begin{aligned}\hat{\boldsymbol{r}} &= \sin\theta\cos\phi\,\boldsymbol{i} + \sin\theta\sin\phi\,\boldsymbol{j} + \cos\theta\,\boldsymbol{k},\\ \hat{\boldsymbol{\theta}} &= \cos\theta\cos\phi\,\boldsymbol{i} + \cos\theta\sin\phi\,\boldsymbol{j} - \sin\theta\,\boldsymbol{k},\\ \hat{\boldsymbol{\phi}} &= -\sin\phi\,\boldsymbol{i} + \cos\phi\,\boldsymbol{j}.\end{aligned}$$

In matrix form, this relationship is written as

$$\begin{pmatrix}\hat{\boldsymbol{r}}\\ \hat{\boldsymbol{\theta}}\\ \hat{\boldsymbol{\phi}}\end{pmatrix} = \begin{pmatrix}\sin\theta\cos\phi & \sin\theta\sin\phi & \cos\theta\\ \cos\theta\cos\phi & \cos\theta\sin\phi & -\sin\theta\\ -\sin\phi & \cos\phi & 0\end{pmatrix}\begin{pmatrix}\boldsymbol{i}\\ \boldsymbol{j}\\ \boldsymbol{k}\end{pmatrix}.$$

The columns (and rows) of the transforming matrix Q are observed to be orthonormal so that Q is an orthogonal matrix. We have $Q^{-1} = Q^{T}$ so that

$$\begin{pmatrix}\boldsymbol{i}\\ \boldsymbol{j}\\ \boldsymbol{k}\end{pmatrix} = \begin{pmatrix}\sin\theta\cos\phi & \cos\theta\cos\phi & -\sin\phi\\ \sin\theta\sin\phi & \cos\theta\sin\phi & \cos\phi\\ \cos\theta & -\sin\theta & 0\end{pmatrix}\begin{pmatrix}\hat{\boldsymbol{r}}\\ \hat{\boldsymbol{\theta}}\\ \hat{\boldsymbol{\phi}}\end{pmatrix};$$

or in expanded form

$$\begin{aligned}\boldsymbol{i} &= \sin\theta\cos\phi\,\hat{\boldsymbol{r}} + \cos\theta\cos\phi\,\hat{\boldsymbol{\theta}} - \sin\phi\,\hat{\boldsymbol{\phi}},\\ \boldsymbol{j} &= \sin\theta\sin\phi\,\hat{\boldsymbol{r}} + \cos\theta\sin\phi\,\hat{\boldsymbol{\theta}} + \cos\phi\,\hat{\boldsymbol{\phi}},\\ \boldsymbol{k} &= \cos\theta\,\hat{\boldsymbol{r}} - \sin\theta\,\hat{\boldsymbol{\theta}}.\end{aligned}$$

2. We need the relationship between the Cartesian and the spherical coordinates, given by

$$x = r\sin\theta\cos\phi, \qquad y = r\sin\theta\sin\phi, \qquad z = r\cos\theta.$$

The Jacobian to compute is

$$\begin{aligned}\begin{vmatrix}\partial x/\partial r & \partial x/\partial\theta & \partial x/\partial\phi\\ \partial y/\partial r & \partial y/\partial\theta & \partial y/\partial\phi\\ \partial z/\partial r & \partial z/\partial\theta & \partial z/\partial\phi\end{vmatrix} &= \begin{vmatrix}\sin\theta\cos\phi & r\cos\theta\cos\phi & -r\sin\theta\sin\phi\\ \sin\theta\sin\phi & r\cos\theta\sin\phi & r\sin\theta\cos\phi\\ \cos\theta & -r\sin\theta & 0\end{vmatrix}\\ &= r^2\sin\theta\begin{vmatrix}\sin\theta\cos\phi & \cos\theta\cos\phi & -\sin\phi\\ \sin\theta\sin\phi & \cos\theta\sin\phi & \cos\phi\\ \cos\theta & -\sin\theta & 0\end{vmatrix}\\ &= r^2\sin\theta\left(\sin^2\theta\cos^2\phi + \cos^2\theta\cos^2\phi + \sin^2\theta\sin^2\phi + \cos^2\theta\sin^2\phi\right)\\ &= r^2\sin\theta\left(\sin^2\theta + \cos^2\theta\right)\left(\sin^2\phi + \cos^2\phi\right) = r^2\sin\theta.\end{aligned}$$

Therefore, $dx\,dy\,dz = r^2 \sin\theta\, dr\, d\theta\, d\phi$.

3. We have

$$\begin{aligned}\int_V f\, dV &= \int_0^{2\pi} \int_0^{\pi} \int_0^{R} f(r) r^2 \sin\theta\, dr\, d\theta\, d\phi \\ &= \int_0^{2\pi} d\phi \int_0^{\pi} \sin\theta\, d\theta \int_0^{R} r^2 f(r)\, dr \\ &= 4\pi \int_0^R r^2 f(r)\, dr,\end{aligned}$$

where we have used $\int_0^{2\pi} d\phi = 2\pi$ and $\int_0^{\pi} \sin\theta\, d\theta = -\cos\theta\big|_0^{\pi} = 2$.

4. To find the mass, we use the result

$$M = \iiint_V \rho(x, y, z)\, dx\, dy\, dz,$$

where ρ is the object's mass density. Here, the density ρ is given by

$$\rho(r) = \rho_0 + (\rho_1 - \rho_0)(r/R),$$

and the total mass of the sphere is given by

$$\begin{aligned}M &= \int_0^{2\pi} \int_0^{\pi} \int_0^{R} \left[\rho_0 + (\rho_1 - \rho_0)(r/R)\right] r^2 \sin\theta\, dr\, d\theta\, d\phi \\ &= 4\pi \int_0^R \left[\rho_0 r^2 + (\rho_1 - \rho_0)\frac{r^3}{R}\right] dr = 4\pi \left[\frac{\rho_0 R^3}{3} + \frac{(\rho_1 - \rho_0)R^3}{4}\right] \\ &= \frac{4}{3}\pi R^3 \left(\frac{1}{4}\rho_0 + \frac{3}{4}\rho_1\right).\end{aligned}$$

The average density of the sphere is its mass divided by its volume, given by

$$\bar{\rho} = \frac{1}{4}\rho_0 + \frac{3}{4}\rho_1.$$

Solutions to the Problems for Lecture 34

1. We begin with

$$\begin{aligned}\hat{\boldsymbol{r}} &= \sin\theta \cos\phi\, \boldsymbol{i} + \sin\theta \sin\phi\, \boldsymbol{j} + \cos\theta\, \boldsymbol{k}, \\ \hat{\boldsymbol{\theta}} &= \cos\theta \cos\phi\, \boldsymbol{i} + \cos\theta \sin\phi\, \boldsymbol{j} - \sin\theta\, \boldsymbol{k}, \\ \hat{\boldsymbol{\phi}} &= -\sin\phi\, \boldsymbol{i} + \cos\phi\, \boldsymbol{j}.\end{aligned}$$

Differentiating,

$$\frac{\partial \hat{\boldsymbol{r}}}{\partial \theta} = \cos\theta\cos\phi\,\boldsymbol{i} + \cos\theta\sin\phi\,\boldsymbol{j} - \sin\theta\,\boldsymbol{k} = \hat{\boldsymbol{\theta}};$$

and

$$\frac{\partial \hat{\boldsymbol{r}}}{\partial \phi} = -\sin\theta\sin\phi\,\boldsymbol{i} + \sin\theta\cos\phi\,\boldsymbol{j} = \sin\theta\,\hat{\boldsymbol{\phi}}.$$

2. The computations are

$$\boldsymbol{\nabla}\cdot\hat{\boldsymbol{r}} = \frac{1}{r^2}\frac{\partial}{\partial r}(r^2) = \frac{2}{r}, \qquad \boldsymbol{\nabla}\times\hat{\boldsymbol{r}} = 0;$$

$$\boldsymbol{\nabla}\cdot\hat{\boldsymbol{\theta}} = \frac{1}{r\sin\theta}\frac{\partial}{\partial\theta}(\sin\theta) = \frac{\cos\theta}{r\sin\theta}, \qquad \boldsymbol{\nabla}\times\hat{\boldsymbol{\theta}} = \frac{\hat{\boldsymbol{\phi}}}{r}\frac{\partial}{\partial r}(r) = \frac{\hat{\boldsymbol{\phi}}}{r};$$

$$\boldsymbol{\nabla}\cdot\hat{\boldsymbol{\phi}} = 0, \qquad \boldsymbol{\nabla}\times\hat{\boldsymbol{\phi}} = \frac{\hat{\boldsymbol{r}}}{r\sin\theta}\frac{\partial}{\partial\theta}(\sin\theta) - \frac{\hat{\boldsymbol{\theta}}}{r}\frac{\partial}{\partial r}(r) = \frac{\hat{\boldsymbol{r}}\cos\theta}{r\sin\theta} - \frac{\hat{\boldsymbol{\theta}}}{r}.$$

3. We use

$$\boldsymbol{\nabla} = \hat{\boldsymbol{r}}\frac{\partial}{\partial r} + \hat{\boldsymbol{\theta}}\frac{1}{r}\frac{\partial}{\partial\theta} + \hat{\boldsymbol{\phi}}\frac{1}{r\sin\theta}\frac{\partial}{\partial\phi},$$

$$\begin{aligned}\boldsymbol{\nabla}\times\boldsymbol{A} = \frac{\hat{\boldsymbol{r}}}{r\sin\theta}\left[\frac{\partial}{\partial\theta}(\sin\theta A_\phi) - \frac{\partial A_\theta}{\partial\phi}\right] &+ \frac{\hat{\boldsymbol{\theta}}}{r}\left[\frac{1}{\sin\theta}\frac{\partial A_r}{\partial\phi} - \frac{\partial}{\partial r}(rA_\phi)\right] \\ &+ \frac{\hat{\boldsymbol{\phi}}}{r}\left[\frac{\partial}{\partial r}(rA_\theta) - \frac{\partial A_r}{\partial\theta}\right],\end{aligned}$$

and the results of Problem 2.

a) If $f = f(r)$, then

$$\begin{aligned}\nabla^2 f = \boldsymbol{\nabla}\cdot\boldsymbol{\nabla}f = \boldsymbol{\nabla}\cdot\left(\frac{df}{dr}\hat{\boldsymbol{r}}\right) &= \frac{d^2 f}{dr^2} + \frac{df}{dr}\boldsymbol{\nabla}\cdot\hat{\boldsymbol{r}} \\ &= \frac{df}{dr^2} + \frac{2}{r}\frac{df}{dr} = \frac{1}{r^2}\frac{d}{dr}\left(r^2\frac{df}{dr}\right).\end{aligned}$$

b) If $\boldsymbol{A} = A_r(r)\hat{\boldsymbol{r}}$, then

$$\nabla^2\boldsymbol{A} = \boldsymbol{\nabla}(\boldsymbol{\nabla}\cdot\boldsymbol{A}) - \boldsymbol{\nabla}\times(\boldsymbol{\nabla}\times\boldsymbol{A}).$$

From the formula for the curl of a vector field in spherical coordinates, one finds $\boldsymbol{\nabla}\times\boldsymbol{A} = 0$. Therefore,

$$\begin{aligned}\nabla^2\boldsymbol{A} = \boldsymbol{\nabla}(\boldsymbol{\nabla}\cdot\boldsymbol{A}) &= \boldsymbol{\nabla}(\boldsymbol{\nabla}\cdot(A_r\hat{\boldsymbol{r}}) \\ &= \boldsymbol{\nabla}\left(\frac{1}{r^2}\frac{d}{dr}(r^2 A_r)\right) = \frac{d}{dr}\left(\frac{1}{r^2}\frac{d}{dr}(r^2 A_r)\right)\hat{\boldsymbol{r}}.\end{aligned}$$

4. Using spherical coordinates, for $r \neq 0$ for which $1/r$ diverges, we have

$$\nabla^2 \left(\frac{1}{r}\right) = \frac{1}{r^2}\frac{\partial}{\partial r}\left(r^2 \frac{\partial}{\partial r}\left(\frac{1}{r}\right)\right) = \frac{1}{r^2}\frac{\partial}{\partial r}(-1) = 0.$$

5. Using spherical coordinates, for $r \neq 0$ for which $1/r^2$ diverges, we have

$$\nabla^2 \left(\frac{\hat{\boldsymbol{r}}}{r^2}\right) = \frac{d}{dr}\left(\frac{1}{r^2}\frac{d}{dr}(\frac{r^2}{r^2})\right)\hat{\boldsymbol{r}} = 0.$$

Solutions to the Practice quiz: Cylindrical and spherical coordinates

1. b. We compute using the Laplacian in cylindrical coordinates:

$$\nabla^2 \left(\frac{1}{\rho}\right) = \frac{1}{\rho}\frac{\partial}{\partial \rho}\rho\frac{\partial}{\partial \rho}\left(\frac{1}{\rho}\right) = \frac{1}{\rho}\frac{\partial}{\partial \rho}\rho\left(-\frac{1}{\rho^2}\right) = -\frac{1}{\rho}\frac{\partial}{\partial \rho}\left(\frac{1}{\rho}\right) = \frac{1}{\rho^3}.$$

2. c. When $\boldsymbol{r} = x\boldsymbol{i}$, the position vector points along the x-axis. Then $\hat{\boldsymbol{r}}$ also points along the x-axis, $\hat{\boldsymbol{\theta}}$ points along the negative z-axis and $\hat{\boldsymbol{\phi}}$ points along the y-axis. We have $(\hat{\boldsymbol{r}}, \hat{\boldsymbol{\theta}}, \hat{\boldsymbol{\phi}}) = (\boldsymbol{i}, -\boldsymbol{k}, \boldsymbol{j})$.

3. c. To find the mass, we use the result

$$M = \iiint_V \rho(x, y, z)\, dx\, dy\, dz,$$

where ρ is the object's mass density. Here, with the density ρ in units of g/cm^3, we have

$$\rho = \rho(r) = 10 - r.$$

The integral is easiest to do in spherical coordinates, and using $dx\, dy\, dz = r^2 \sin\theta\, dr\, d\theta\, d\phi$, and computing in grams, we have

$$\begin{aligned} M &= \int_0^{2\pi}\int_0^{\pi}\int_0^5 (10 - r)\, r^2 \sin\theta\, dr\, d\theta\, d\phi \\ &= 4\pi \int_0^5 (10r^2 - r^3)\, dr \\ &= 4\pi \left(\frac{10}{3}r^3 - \frac{1}{4}r^4\right)\Bigg|_0^5 = \frac{3125\pi}{3}\,\text{g} \\ &\approx 3272\,\text{g} \approx 3.3\,\text{kg}. \end{aligned}$$

Solutions to the Problems for Lecture 35

1. We parameterize a circle of radius R by

$$x(\theta) = R\cos\theta, \qquad y(\theta) = R\sin\theta,$$

where the angle θ goes from 0 to 2π. The infinitesimal arc length ds is given by

$$\begin{aligned} ds &= \sqrt{(dx)^2 + (dy)^2} = \sqrt{x'(\theta)^2 + y'(\theta)^2}\, d\theta \\ &= \sqrt{R^2\sin^2\theta + R^2\cos^2\theta}\, d\theta = R\, d\theta. \end{aligned}$$

The circumference of a circle — or perimeter P — is then given by the line integral

$$P = \int_C ds = \int_0^{2\pi} R d\theta = 2\pi R.$$

2. Place the semi-circular wire in the upper half of the x-y plane. Since arc length is given by $R\Delta\theta$, the mass density of the wire increases linearly with the polar angle θ. Then in polar coordinates,

$$\lambda(\theta) = \lambda_0 + \frac{1}{\pi}(\lambda_1 - \lambda_0)\theta.$$

To calculate the mass of the wire, we again parameterize the semi-circle of radius R by

$$x(\theta) = R\cos\theta, \qquad y(\theta) = R\sin\theta,$$

where now the angle θ goes from 0 to π. The infinitesimal arc length ds is given by $ds = R\, d\theta$, and the total mass of the wire is given by

$$\begin{aligned} M &= \int_C \lambda\, ds = \int_0^{\pi} \left(\lambda_0 + \frac{1}{\pi}(\lambda_1 - \lambda_0)\theta\right) R\, d\theta \\ &= R\left(\lambda_0\pi + (\lambda_1 - \lambda_0)\frac{\pi}{2}\right) = \pi R\frac{(\lambda_0 + \lambda_1)}{2}, \end{aligned}$$

which is the length of the wire πR times the average linear mass density.

Solutions to the Problems for Lecture 36

1. We start with the exact integral formula for the perimeter of an ellipse, and Taylor series expand the integrand in the eccentricity e, keeping only the first two terms. Using the Taylor series approximation, $\sqrt{1+\delta} \approx 1 + \delta/2$, we have

$$P = 4a \int_0^{\pi/2} \sqrt{1 - e^2 \cos^2\theta}\, d\theta \approx 4a \int_0^{\pi/2} \left(1 - \frac{1}{2}e^2\cos^2\theta\right) d\theta$$

$$= 4a\left(\frac{\pi}{2} - \frac{1}{2}e^2 \int_0^{\pi/2} \cos^2\theta\, d\theta\right) = 4a\left(\frac{\pi}{2} - \frac{\pi}{8}e^2\right) = 2\pi a\left(1 - \frac{1}{4}e^2\right),$$

which is slightly less than the perimeter of a circle of radius a.

Solutions to the Problems for Lecture 37

1. We integrate $\boldsymbol{u} = -y\boldsymbol{i} + x\boldsymbol{j}$ counterclockwise around the square. We write

$$\oint_C \boldsymbol{u} \cdot d\boldsymbol{r} = \int_{C_1} \boldsymbol{u} \cdot d\boldsymbol{r} + \int_{C_2} \boldsymbol{u} \cdot d\boldsymbol{r} + \int_{C_3} \boldsymbol{u} \cdot d\boldsymbol{r} + \int_{C_4} \boldsymbol{u} \cdot d\boldsymbol{r},$$

where the curves C_i represent the four sides of the square. On C_1 from $(0,0)$ to $(L,0)$, we have $y = 0$ and $d\boldsymbol{r} = dx\boldsymbol{i}$ so that $\int_{C_1} \boldsymbol{u} \cdot d\boldsymbol{r} = 0$. On C_2 from $(L,0)$ to (L,L), we have $x = L$ and $d\boldsymbol{r} = dy\boldsymbol{j}$ so that $\int_{C_2} \boldsymbol{u} \cdot d\boldsymbol{r} = \int_0^L L dy = L^2$. On C_3 from (L,L) to $(0,L)$, we have $y = L$ and $d\boldsymbol{r} = dx\boldsymbol{i}$ so that $\int_{C_3} \boldsymbol{u} \cdot d\boldsymbol{r} = \int_L^0 -L dx = L^2$. The sign of this term is tricky, but notice that the curve is going in the -$\boldsymbol{i}$ direction and so is the x-component of the vector field so the dot product should be positive. On C_4 from $(0,L)$ to $(0,0)$, we have $x = 0$ and $d\boldsymbol{r} = dy\boldsymbol{j}$ so that $\int_{C_4} \boldsymbol{u} \cdot d\boldsymbol{r} = 0$. Summing the four contributions, we found

$$\oint_C \boldsymbol{u} \cdot d\boldsymbol{r} = 2L^2,$$

which is twice the area of the square.

2. We integrate $\boldsymbol{u} = -y\boldsymbol{i} + x\boldsymbol{j}$ counterclockwise around a unit circle. To parameterize a circle with radius R, we write

$$x = R\cos\theta, \qquad y = R\sin\theta.$$

Therefore, $\boldsymbol{u} = -R\sin\theta\boldsymbol{i} + R\cos\theta\boldsymbol{j}$ and $d\boldsymbol{r} = (-R\sin\theta\boldsymbol{i} + R\cos\theta\boldsymbol{j})d\theta$. We have $\boldsymbol{u}\cdot d\boldsymbol{r} = R^2 d\theta$ and

$$\oint_C \boldsymbol{u}\cdot d\boldsymbol{r} = \int_0^{2\pi} R^2\,d\theta = 2\pi R^2,$$

which is twice the area of the circle.

Solutions to the Problems for Lecture 38

1. We define our coordinate system with the x-axis pointing downward and the origin at the initial position of the mass. With

$$\boldsymbol{F} = mg\boldsymbol{i}, \qquad d\boldsymbol{r} = dx\boldsymbol{i},$$

the work done by gravity as the mass falls a distance h is given by

$$W = \int_C \boldsymbol{F}\cdot d\boldsymbol{r} = \int_0^h mg\,dx = mgh.$$

With $\boldsymbol{v}_f$ the final velocity of the mass, and with the initial velocity equal to zero, we have from the work-energy theorem,

$$mgh = \frac{1}{2}m|\boldsymbol{v}_f|^2,$$

or

$$|\boldsymbol{v}_f| = \sqrt{2gh}.$$

Solutions to the Practice quiz: Line integrals

1. d. We have

$$ds = \sqrt{(dx)^2 + (dy)^2} = \sqrt{1 + (dy/dx)^2}\,dx = \sqrt{1 + (2x)^2}\,dx.$$

Therefore, the arc length is given by

$$\int_0^1 \sqrt{1+4x^2}\,dx.$$

2. c. We integrate $\boldsymbol{u} = -y\boldsymbol{i} + x\boldsymbol{j}$ counterclockwise around the right triangle. We write

$$\oint_C \boldsymbol{u}\cdot d\boldsymbol{r} = \int_{C_1} \boldsymbol{u}\cdot d\boldsymbol{r} + \int_{C_2} \boldsymbol{u}\cdot d\boldsymbol{r} + \int_{C_3} \boldsymbol{u}\cdot d\boldsymbol{r},$$

where the curves C_i represent the three sides of the triangle. On C_1 from $(0,0)$ to $(L,0)$, we have $y = 0$ and $d\boldsymbol{r} = dx\boldsymbol{i}$ so that $\int_{C_1} \boldsymbol{u} \cdot d\boldsymbol{r} = 0$. On C_2 from $(L,0)$ to $(0,L)$, we parameterize the line segment by $\boldsymbol{r} = (L-s)\boldsymbol{i} + s\boldsymbol{j}$ as s goes from zero to L so that $d\boldsymbol{r} = -ds\boldsymbol{i} + ds\boldsymbol{j}$. Therefore, on this line segment, $\boldsymbol{u} \cdot d\boldsymbol{r} = (x+y)ds = L\,ds$. We have $\int_{C_2} \boldsymbol{u} \cdot d\boldsymbol{r} = \int_0^L L\,ds = L^2$. On C_3 from $(0,L)$ to $(0,0)$, we have $x = 0$ and $d\boldsymbol{r} = dy\boldsymbol{j}$ so that $\int_{C_4} \boldsymbol{u} \cdot d\boldsymbol{r} = 0$. Summing the three contributions, we found

$$\oint_C \boldsymbol{u} \cdot d\boldsymbol{r} = L^2,$$

which is twice the area of the triangle.

3. a. Define the x-axis to point vertically upward. The gravitational force is given by $\mathbf{F} = -mg\mathbf{i}$ and the work done by gravity on the way up is $-mgx_{\text{max}}$ and the work done by gravity on the way down is mgx_{max}, where x_{max} is the maximum height attained by the mass. The total work done is zero.

Solutions to the Problems for Lecture 39

1.

a) The unrolled cylinder is a rectangle with dimensions as shown on the figure below:

The lateral surface area is $A = 2\pi ab$.

b) Define the cylinder parametrically as

$$\boldsymbol{r} = a\cos\theta\,\boldsymbol{i} + a\sin\theta\,\boldsymbol{j} + z\,\boldsymbol{k}, \qquad \text{for } 0 \leq z \leq b \quad \text{and} \quad 0 \leq \theta \leq 2\pi.$$

To find the infinitesimal surface element, we compute the partial derivatives of $\boldsymbol{r}$:

$$\frac{\partial \boldsymbol{r}}{\partial \theta} = -a\sin\theta\,\boldsymbol{i} + a\cos\theta\,\boldsymbol{j}, \qquad \frac{\partial \boldsymbol{r}}{\partial z} = \boldsymbol{k}.$$

The cross product is

$$\frac{\partial \boldsymbol{r}}{\partial \theta} \times \frac{\partial \boldsymbol{r}}{\partial z} = \begin{vmatrix} \boldsymbol{i} & \boldsymbol{j} & \boldsymbol{k} \\ -a\sin\theta & a\cos\theta & 0 \\ 0 & 0 & 1 \end{vmatrix} = a\cos\theta\,\boldsymbol{i} + a\sin\theta\,\boldsymbol{j},$$

so that

$$\left|\frac{\partial \boldsymbol{r}}{\partial \theta} \times \frac{\partial \boldsymbol{r}}{\partial z}\right| = \sqrt{a^2\cos^2\theta + a^2\sin^2\theta} = a.$$

The surface area is given by

$$A = \int_S dS = \int_0^b \int_0^{2\pi} a\,d\theta\,dz = 2\pi ab.$$

2.

a) The unrolled cone is a circular sector with dimensions as shown on the figure below:

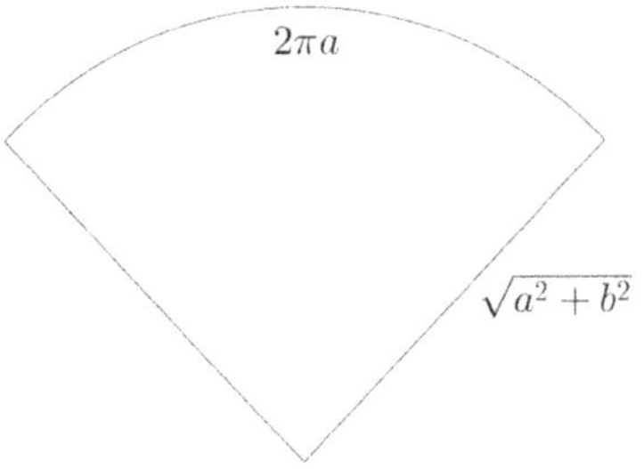

The lateral surface area is the sector of a circle of radius $\sqrt{a^2+b^2}$. Its area is found from

$$\begin{aligned} A &= \frac{\text{arc length of sector}}{\text{circumference of circle}} \times \text{area of circle} \\ &= \frac{2\pi a}{2\pi\sqrt{a^2+b^2}} \times \pi(a^2+b^2) \\ &= \pi a\sqrt{a^2+b^2} = \pi ab\sqrt{1+\left(\frac{a}{b}\right)^2}. \end{aligned}$$

b) Define the cone parametrically as

$$\boldsymbol{r} = \frac{az}{b}\cos\theta\,\boldsymbol{i} + \frac{az}{b}\sin\theta\,\boldsymbol{j} + z\,\boldsymbol{k}, \qquad \text{for } 0 \le z \le b \quad \text{and} \quad 0 \le \theta \le 2\pi,$$

To find the infinitesimal surface element, we compute the partial derivatives of $\boldsymbol{r}$:

$$\frac{\partial \boldsymbol{r}}{\partial \theta} = -\frac{az}{b}\sin\theta\,\boldsymbol{i} + \frac{az}{b}\cos\theta\,\boldsymbol{j}, \qquad \frac{\partial \boldsymbol{r}}{\partial z} = \frac{a}{b}\cos\theta\,\boldsymbol{i} + \frac{a}{b}\sin\theta + \boldsymbol{k}.$$

The cross product is

$$\frac{\partial \boldsymbol{r}}{\partial \theta} \times \frac{\partial \boldsymbol{r}}{\partial z} = \begin{vmatrix} \boldsymbol{i} & \boldsymbol{j} & \boldsymbol{k} \\ -\frac{az}{b}\sin\theta & \frac{az}{b}\cos\theta & 0 \\ \frac{a}{b}\cos\theta & \frac{a}{b}\sin\theta & 1 \end{vmatrix} = \frac{az}{b}\cos\theta\,\boldsymbol{i} + \frac{az}{b}\sin\theta\,\boldsymbol{j} - \frac{a^2 z}{b^2}\,\boldsymbol{k},$$

so that

$$\begin{aligned} \left|\frac{\partial \boldsymbol{r}}{\partial \theta} \times \frac{\partial \boldsymbol{r}}{\partial z}\right| &= \sqrt{\frac{a^2z^2}{b^2}\cos^2\theta + \frac{a^2z^2}{b^2}\sin^2\theta + \frac{a^4z^2}{b^4}} \\ &= \frac{az}{b}\sqrt{1 + \frac{a^2}{b^2}}. \end{aligned}$$

The surface area is given by

$$\begin{aligned} A = \int_S dS &= \int_0^b \int_0^{2\pi} \frac{az}{b}\sqrt{1 + \frac{a^2}{b^2}}\, d\theta\, dz \\ &= \frac{a}{b}\sqrt{1 + \frac{a^2}{b^2}} \int_0^b z\, dz \int_0^{2\pi} d\theta \\ &= \pi ab\sqrt{1 + \left(\frac{a}{b}\right)^2}. \end{aligned}$$

Solutions to the Problems for Lecture 40

1. For the paraboloid, we have

$$z(x,y) = \frac{b}{a^2}\left(x^2 + y^2\right)$$

The surface area is given by

$$S = \int_S dS = \int_S \sqrt{1 + \left(\frac{\partial z}{\partial x}\right)^2 + \left(\frac{\partial z}{\partial y}\right)^2}\, dx\, dy.$$

Here,

$$\frac{\partial z}{\partial x} = \frac{2bx}{a^2}, \qquad \frac{\partial z}{\partial y} = \frac{2by}{a^2},$$

so that

$$\begin{aligned} S &= \int_S \sqrt{1 + 4b^2x^2/a^4 + 4b^2y^2/a^4}\, dx\, dy \\ &= \frac{2b}{a^2} \int_S \sqrt{\frac{a^4}{4b^2} + (x^2 + y^2)}\, dx\, dy. \end{aligned}$$

We integrate in polar coordinates. Let

$$x = r\cos\theta, \qquad y = r\sin\theta,$$

and $dx\, dy = r\, dr\, d\theta$. The integral becomes

$$\begin{aligned} S &= \frac{2b}{a^2} \int_0^{2\pi} \int_0^a \sqrt{\frac{a^4}{4b^2} + r^2}\; r\, dr\, d\theta \\ &= \frac{2\pi b}{a^2} \int_{a^4/4b^2}^{a^2 + a^4/4b^2} u^{1/2}\, du \\ &= \frac{4\pi b}{3a^2} \left(\left(a^2 + \frac{a^4}{4b^2} \right)^{3/2} - \left(\frac{a^4}{4b^2} \right)^{3/2} \right) \\ &= \frac{4}{3}\pi ab \left(\left(1 + \frac{a^2}{4b^2} \right)^{3/2} - \left(\frac{a}{2b} \right)^3 \right). \end{aligned}$$

Solutions to the Problems for Lecture 41

1. Use cylindrical coordinates with the origin at the exact center of the cylinder, with the z-axis down the long symmetry axis. The flat top disk of the cylinder will be at $z = l/2$ and the bottom disk of the cylinder will be at $z = -l/2$. The lateral curved surface of the cylinder is at $\rho = a$.

We have for the surface integral on the lateral surface,

$$\boldsymbol{r} = a\hat{\boldsymbol{\rho}} + z\hat{\boldsymbol{z}}, \qquad d\boldsymbol{S} = \hat{\boldsymbol{\rho}} dS,$$

and for the total surface area of the lateral surface, $S = 2\pi al$. Therefore,

$$\int_S \boldsymbol{r} \cdot d\boldsymbol{S} = a \int_S dS = 2\pi a^2 l.$$

On the top disk of the cylinder, we have

$$\boldsymbol{r} = \rho\hat{\boldsymbol{\rho}} + \frac{l}{2}\hat{\boldsymbol{z}}, \qquad d\boldsymbol{S} = \hat{\boldsymbol{z}}\, dS,$$

and for the total surface area of the top disk, $S = \pi a^2$. Therefore,

$$\int_S \boldsymbol{r} \cdot d\boldsymbol{S} = \frac{l}{2} \int_S dS = \frac{\pi a^2 l}{2}.$$

The bottom disk of the cylinder will yield the same result, so that

$$\oint_S \boldsymbol{r} \cdot d\boldsymbol{S} = 2\pi a^2 l + 2 \times \frac{\pi a^2 l}{2} = 3\pi a^2 l,$$

which is three times the volume of the cylinder.

Solutions to the Problems for Lecture 42

1. To find the mass flux, we use

$$\boldsymbol{u}(\boldsymbol{r}) = u_{\mathrm{m}} \left(1 - \left(\frac{r}{R}\right)^2\right) \boldsymbol{k}.$$

With $d\boldsymbol{S} = \boldsymbol{k}\, r\, dr\, d\theta$, the mass flux through a cross section of the pipe is given by

$$\begin{aligned}\int_S \rho \boldsymbol{u} \cdot d\boldsymbol{S} &= \int_0^{2\pi} \int_0^R \rho u_{\mathrm{m}} \left(1 - \left(\frac{r}{R}\right)^2\right) r\, dr\, d\theta \\ &= 2\pi R^2 \rho u_{\mathrm{m}} \int_0^1 \left(1 - s^2\right) s\, ds \\ &= \frac{1}{2} \pi R^2 \rho u_{\mathrm{m}}.\end{aligned}$$

This is one-half of what it would be if the entire fluid was moving with velocity u_{m}. Using the formula for u_m, we have

$$\int_S \rho \boldsymbol{u} \cdot d\boldsymbol{S} = \frac{\pi G R^4}{8\nu}.$$

This result is called the Hagen-Poiseuille equation, which relates the pressure gradient to the radius of the pipe for a fixed mass flux.

Solutions to the Practice quiz: Surface integrals

1. d. The parameterization of the torus is given by

$$x = (R + r\cos\theta)\cos\phi, \qquad y = (R + r\cos\theta)\sin\phi, \qquad z = r\sin\theta,$$

so that the position vector is given by

$$\mathbf{r} = (R + r\cos\theta)\cos\phi\,\mathbf{i} + (R + r\cos\theta)\sin\phi\,\mathbf{j} + r\sin\theta\,\mathbf{k}.$$

The partial derivatives are

$$\begin{aligned}\frac{\partial \mathbf{r}}{\partial \theta} &= -r\sin\theta\cos\phi\,\mathbf{i} - r\sin\theta\sin\phi\,\mathbf{j} + r\cos\theta\,\mathbf{k},\\ \frac{\partial \mathbf{r}}{\partial \phi} &= -(R + r\cos\theta)\sin\phi\,\mathbf{i} + (R + r\cos\theta)\cos\phi\,\mathbf{j},\end{aligned}$$

and

$$\begin{aligned}\frac{\partial \mathbf{r}}{\partial \theta} \times \frac{\partial \mathbf{r}}{\partial \phi} &= \begin{vmatrix} \mathbf{i} & \mathbf{j} & \mathbf{k} \\ -r\sin\theta\cos\phi & -r\sin\theta\sin\phi & r\cos\theta \\ -(R + r\sin\theta)\sin\phi & (R + r\cos\theta)\cos\phi & 0 \end{vmatrix}\\ &= -r(R + r\cos\theta)\cos\theta\cos\phi\,\mathbf{i} - r(R + r\cos\theta)\cos\theta\sin\phi\,\mathbf{j}\\ &\qquad - r(R + r\cos\theta)\sin\theta\,\mathbf{k}.\end{aligned}$$

Therefore,

$$\begin{aligned}&\left|\frac{\partial \mathbf{r}}{\partial \theta} \times \frac{\partial \mathbf{r}}{\partial \phi}\right|\\ &= \left[r^2(R + r\cos\theta)^2\cos^2\theta(\cos^2\phi + \sin^2\phi) + r^2(R + r\cos\theta)^2\sin^2\theta\right]^{1/2}\\ &\qquad = \left[r^2(R + r\cos\theta)^2(\cos^2\theta + \sin^2\theta)\right]^{1/2} = r(R + r\cos\theta),\end{aligned}$$

and

$$dS = r(R + r\cos\theta)\,d\theta\,d\phi.$$

2. c. In cylindrical coordinates, $\boldsymbol{u} = x\boldsymbol{i} + y\boldsymbol{j} = \rho\hat{\boldsymbol{\rho}}$. The cylinder ends have normal vectors $\hat{\boldsymbol{z}}$ and $-\hat{\boldsymbol{z}}$, which are perpendicular to $\boldsymbol{u}$. On the side of the cylinder, we have $\boldsymbol{u} = R\hat{\boldsymbol{\rho}}$ and $d\boldsymbol{S} = \hat{\boldsymbol{\rho}}dS$, so that

$$\oint_S \boldsymbol{u}\cdot d\boldsymbol{S} = R\int dS = R(2\pi RL) = 2\pi R^2 L.$$

3. a. We perform the flux integral in spherical coordinates. On the surface of the sphere of radius R, we have

$$\boldsymbol{u} = z\boldsymbol{k} = (R\cos\theta)(\cos\theta\hat{\boldsymbol{r}} - \sin\theta\hat{\boldsymbol{\theta}}),$$

and

$$dS = \hat{\boldsymbol{r}}\, R^2 \sin\theta \, d\theta \, d\phi.$$

Therefore, the surface integral over the upper hemisphere becomes

$$\begin{aligned}\int_S \boldsymbol{u} \cdot d\boldsymbol{S} &= \int_0^{2\pi} \int_0^{\pi/2} R^3 \cos^2\theta \sin\theta \, d\theta \, d\phi \\ &= 2\pi R^3 \int_0^{\pi/2} \cos^2\theta \sin\theta \, d\theta = 2\pi R^3 \int_0^1 w^2 \, dw = \frac{2\pi}{3} R^3.\end{aligned}$$

Solutions to the Problems for Lecture 43

1. With $\phi(\boldsymbol{r}) = x^2y + xy^2 + z$:

a) $\boldsymbol{\nabla}\phi = (2xy + y^2)\boldsymbol{i} + (x^2 + 2xy)\boldsymbol{j} + \boldsymbol{k}$

b) Using the gradient theorem, $\int_C \boldsymbol{\nabla}\phi \cdot d\boldsymbol{r} = \phi(1,1,1) - \phi(0,0,0) = 3$.

c) Integrating over the three directed line segments given by (1) $(0,0,0)$ to $(1,0,0)$; (2) $(1,0,0)$ to $(1,1,0)$, and; (3) $(1,1,0)$ to $(1,1,1)$:

$$\begin{aligned}\int_C \boldsymbol{\nabla}\phi \cdot d\boldsymbol{r} &= \int_{C_1} \boldsymbol{\nabla}\phi \cdot d\boldsymbol{r} + \int_{C_2} \boldsymbol{\nabla}\phi \cdot d\boldsymbol{r} + \int_{C_3} \boldsymbol{\nabla}\phi \cdot d\boldsymbol{r} \\ &= 0 + \int_0^1 (1 + 2y)\, dy + \int_0^1 dz \\ &= 3.\end{aligned}$$

Solutions to the Problems for Lecture 44

1. With $\boldsymbol{u} = (2xy + z^2)\boldsymbol{i} + (2yz + x^2)\boldsymbol{j} + (2zx + y^2)\boldsymbol{k}$:

a)

$$\begin{aligned}\boldsymbol{\nabla} \times \boldsymbol{u} &= \begin{vmatrix} \boldsymbol{i} & \boldsymbol{j} & \boldsymbol{k} \\ \partial/\partial x & \partial/\partial y & \partial/\partial z \\ 2xy + z^2 & 2yz + x^2 & 2zx + y^2 \end{vmatrix} \\ &= (2y - 2y)\boldsymbol{i} + (2z - 2z)\boldsymbol{j} + (2x - 2x)\boldsymbol{k} \\ &= 0.\end{aligned}$$

b) We need to satisfy

$$\frac{\partial \phi}{\partial x} = 2xy + z^2, \qquad \frac{\partial \phi}{\partial y} = 2yz + x^2, \qquad \frac{\partial \phi}{\partial z} = 2zx + y^2.$$

Integrate the first equation to get

$$\phi = \int (2xy + z^2)\, dx = x^2 y + xz^2 + f(y, z).$$

Take the derivative with respect to y and satsify the second equation:

$$x^2 + \frac{\partial f}{\partial y} = 2yz + x^2 \qquad \text{or} \qquad \frac{\partial f}{\partial y} = 2yz.$$

Integrate this equation for f to get

$$f = \int 2yz\, dy = y^2 z + g(z).$$

Take the derivative of $\phi = x^2y + xz^2 + y^2z + g(z)$ with respect to z and satisfy the last gradient equation:

$$2xz + y^2 + g'(z) = 2zx + y^2 \qquad \text{or} \qquad g'(z) = 0.$$

Therefore, $g(z) = c$ where c is a constant, and $\phi = x^2y + y^2z + z^2x + c$.

Solutions to the Problems for Lecture 45

1. Let v_{escape} be the magnitude of the escape velocity for a mass launched perpendicular to the Earth's surface. When the mass reaches infinity, its velocity should be exactly zero. Conservation of energy results in

$$\frac{1}{2} m v_{\text{escape}}^2 - G\frac{mM}{R} = 0,$$

or

$$v_{\text{escape}}^2 = 2\frac{GM}{R^2} R = 2gR.$$

Therefore, we have

$$v_{\text{escape}} = \sqrt{2gR}.$$

Solutions to the Practice quiz: Gradient theorem

1. b. $\int_C \boldsymbol{\nabla}\phi \cdot d\boldsymbol{r} = \phi(1,1,1) - \phi(0,0,0) = 1.$

2. a. Since $\boldsymbol{\nabla} \times \boldsymbol{u} = \boldsymbol{\nabla} \times (y\boldsymbol{i} + x\boldsymbol{j}) = 0$, the line integral $\boldsymbol{u}$ around any closed curve is zero. By inspection, we can also observe that $\boldsymbol{u} = \boldsymbol{\nabla}\phi$, where $\phi = xy + c$.

3. c. To solve the multiple choice question, we can always take the gradients of the four choices. Without the advantage of multiple choice, however, we need to compute ϕ and we do so here. We solve

$$\frac{\partial \phi}{\partial x} = 2x + y, \qquad \frac{\partial \phi}{\partial y} = 2y + x, \qquad \frac{\partial \phi}{\partial z} = 1.$$

Integrating the first equation with respect to x holding y and z fixed, we find

$$\phi = \int (2x + y)\, dx = x^2 + xy + f(y, z).$$

Differentiating ϕ with respect to y and using the second equation, we obtain

$$x + \frac{\partial f}{\partial y} = 2y + x \qquad \text{or} \qquad \frac{\partial f}{\partial y} = 2y.$$

Another integration results in $f(y, z) = y^2 + g(z)$. Finally, differentiating ϕ with respect to z yields $g'(z) = 1$, or $g(z) = z + c$. The final solution is

$$\phi(x, y, z) = x^2 + xy + y^2 + z + c.$$

Answer c. is correct with the constant $c = 0$.

Solutions to the Problems for Lecture 46

1. Using spherical coordinates, let $\boldsymbol{u} = u_r(r, \theta, \phi)\hat{\boldsymbol{r}} + u_\theta(r, \theta, \phi)\hat{\boldsymbol{\theta}} + u_\phi(r, \theta, \phi)\hat{\boldsymbol{\phi}}$. Then the volume integral becomes

$$\int_V (\boldsymbol{\nabla} \cdot \boldsymbol{u})\, dV$$
$$= \int_0^{2\pi} \int_0^{\pi} \int_0^{R} \left(\frac{1}{r^2}\frac{\partial}{\partial r}(r^2 u_r) + \frac{1}{r\sin\theta}\frac{\partial}{\partial \theta}(\sin\theta\, u_\theta) + \frac{1}{r\sin\theta}\frac{\partial u_\phi}{\partial \phi} \right) r^2 \sin\theta\, dr\, d\theta\, d\phi.$$

Each term in the integrand can be integrated once. The first term is integrated as

$$\int_0^{2\pi} \int_0^{\pi} \int_0^{R} \left(\frac{1}{r^2}\frac{\partial}{\partial r}(r^2 u_r) \right) r^2 \sin\theta\, dr\, d\theta\, d\phi$$
$$= \int_0^{2\pi} \int_0^{\pi} \left(\int_0^{R} \frac{\partial}{\partial r}(r^2 u_r)\, dr \right) \sin\theta\, d\theta\, d\phi = \int_0^{2\pi} \int_0^{\pi} u_r(R, \theta, \phi) R^2 \sin\theta\, d\theta\, d\phi.$$

The second term is integrated as

$$\int_0^{2\pi}\int_0^{\pi}\int_0^R \left(\frac{1}{r\sin\theta}\frac{\partial}{\partial\theta}(\sin\theta\, u_\theta)\right) r^2 \sin\theta\, dr\, d\theta\, d\phi$$
$$= \int_0^{2\pi}\int_0^R \left(\int_0^{\pi} \frac{\partial}{\partial\theta}(\sin\theta\, u_\theta)\, d\theta\right) r\, dr\, d\phi$$
$$= \int_0^{2\pi}\int_0^R (\sin(\pi)\, u_\theta(r,\pi,\phi) - \sin(0)\, u_\theta(r,0,\phi))\, r\, dr\, d\phi = 0,$$

since $\sin(\pi) = \sin(0) = 0$. Similarly, the third term is integrated as

$$\int_0^{2\pi}\int_0^{\pi}\int_0^R \left(\frac{1}{r\sin\theta}\frac{\partial u_\phi}{\partial\phi}\right) r^2 \sin\theta\, dr\, d\theta\, d\phi$$
$$= \int_0^{\pi}\int_0^R \left(\int_0^{2\pi} \frac{\partial u_\phi}{\partial\phi}\, d\phi\right) r\, dr\, d\theta$$
$$= \int_0^{\pi}\int_0^R (u_\phi(r,\theta,2\pi) - u_\phi(r,\theta,0))\, r\, dr\, d\theta = 0,$$

since $u_\phi(r,\theta,2\pi) = u_\phi(r,\theta,0)$ because ϕ is a periodic variable with the same physical location at 0 and 2π.

Therefore, we have

$$\int_V (\boldsymbol{\nabla}\cdot\boldsymbol{u})\, dV = \int_0^{2\pi}\int_0^{\pi} u_r(R,\theta,\phi) R^2 \sin\theta\, d\theta\, d\phi = \oint_S \boldsymbol{u}\cdot d\boldsymbol{S},$$

where S is a sphere of radius R located at the origin, with unit normal vector given by $\hat{\boldsymbol{r}}$, and infinitesimal surface area given by $dS = R^2 \sin\theta\, d\theta\, d\phi$.

Solutions to the Problems for Lecture 47

1. With $\boldsymbol{u} = x^2 y\,\boldsymbol{i} + y^2 z\,\boldsymbol{j} + z^2 x\,\boldsymbol{k}$, we use $\boldsymbol{\nabla}\cdot\boldsymbol{u} = 2xy + 2yz + 2zx$. We have for the left-hand side of the divergence theorem,

$$\int_V (\boldsymbol{\nabla}\cdot\boldsymbol{u})\, dV = 2\int_0^L\int_0^L\int_0^L (xy + yz + zx)\, dx\, dy\, dz$$
$$= 2\left[\int_0^L x\, dx \int_0^L y\, dy \int_0^L dz + \int_0^L dx \int_0^L y\, dy \int_0^L z\, dz \right.$$
$$\left. + \int_0^L x\, dx \int_0^L dy \int_0^L z\, dz\right] = 2(L^5/4 + L^5/4 + L^5/4) = 3L^5/2.$$

For the right-hand side of the divergence theorem, the flux integral only has nonzero contributions from the three sides located at $x = L$, $y = L$ and $z = L$. The corresponding unit normal vectors are $\boldsymbol{i}$, $\boldsymbol{j}$ and $\boldsymbol{k}$, and the corresponding integrals are

$$\begin{aligned}\oint_S \boldsymbol{u} \cdot d\boldsymbol{S} &= \int_0^L \int_0^L L^2 y \, dy \, dz + \int_0^L \int_0^L L^2 z \, dx \, dz + \int_0^L \int_0^L L^2 x \, dx \, dy \\ &= L^5/2 + L^5/2 + L^5/2 \\ &= 3L^5/2.\end{aligned}$$

2. With $\boldsymbol{r} = x\boldsymbol{i} + y\boldsymbol{j} + z\boldsymbol{k}$, we have $\boldsymbol{\nabla} \cdot \boldsymbol{r} = 3$. Therefore, from the divergence theorem we have

$$\int_S \boldsymbol{r} \cdot d\boldsymbol{S} = \int_V \boldsymbol{\nabla} \cdot \boldsymbol{r} \, dV = 3 \int_V dV = 3L^3.$$

Note that the integral is equal to three times the volume of the box and is independent of the placement and orientation of the coordinate system.

Solutions to the Problems for Lecture 48

1. With $\boldsymbol{u} = \hat{\boldsymbol{r}}/r$, we use spherical coordinates to compute

$$\boldsymbol{\nabla} \cdot \boldsymbol{u} = \frac{1}{r^2} \frac{d}{dr}(r) = \frac{1}{r^2}.$$

Therefore, for the left-hand side of the divergence theorem we have

$$\int_V (\boldsymbol{\nabla} \cdot \boldsymbol{u}) \, dV = 4\pi \int_0^R \left(\frac{1}{r^2}\right) r^2 \, dr = 4\pi R.$$

For the right-hand side of the divergence theorem, we have for a sphere of radius R centered at the origin, $d\boldsymbol{S} = \hat{\boldsymbol{r}} \, dS$ and

$$\oint_S \boldsymbol{u} \cdot d\boldsymbol{S} = \oint_S \frac{1}{R} \, dS = \frac{4\pi R^2}{R} = 4\pi R.$$

2. With $\boldsymbol{r} = x\boldsymbol{i} + y\boldsymbol{j} + z\boldsymbol{k}$, we have $\boldsymbol{\nabla} \cdot \boldsymbol{r} = 3$. Therefore, from the divergence theorem we have

$$\int_S \boldsymbol{r} \cdot d\boldsymbol{S} = \int_V \boldsymbol{\nabla} \cdot \boldsymbol{r} \, dV = 3 \int_V dV = 3\left(\frac{4}{3}\pi R^3\right) = 4\pi R^3.$$

Note that the integral is equal to three times the volume of the sphere and is independent of the placement and orientation of the coordinate system.

3. We consider the velocity field given by

$$\boldsymbol{u}(x, y, z) = \frac{\Lambda(x\boldsymbol{i} + y\boldsymbol{j} + z\boldsymbol{k})}{4\pi(x^2 + y^2 + z^2)^{3/2}}.$$

a) Using spherical coordinates, we have $\boldsymbol{r} = r\hat{\boldsymbol{r}}$ with $r = \sqrt{x^2 + y^2 + z^2}$. Therefore,

$$\boldsymbol{u} = \frac{\Lambda\hat{\boldsymbol{r}}}{4\pi r^2}.$$

b) We compute $\boldsymbol{\nabla} \cdot \boldsymbol{u}$ for $r \neq 0$ using spherical coordinates:

$$\boldsymbol{\nabla} \cdot \boldsymbol{u} = \frac{1}{r^2}\frac{\partial}{\partial r}\left(r^2\left(\frac{\Lambda}{4\pi r^2}\right)\right) = 0.$$

c) We now consider the volume integral of $\boldsymbol{\nabla} \cdot \boldsymbol{u}$, i.e.,

$$\int_V \boldsymbol{\nabla} \cdot \boldsymbol{u}\, dV.$$

If V does not contain the origin, then this volume integral is zero. If V contains the origin, and since $\boldsymbol{\nabla} \cdot \boldsymbol{u} = 0$ everywhere except at the origin, we need only integrate over a small sphere of volume $V' \in V$ centered at the origin. We therefore have from the divergence theorem

$$\int_V \boldsymbol{\nabla} \cdot \boldsymbol{u}\, dV = \int_{V'} \boldsymbol{\nabla} \cdot \boldsymbol{u}\, dV = \oint_{S'} \boldsymbol{u} \cdot d\boldsymbol{S},$$

where the surface S' is now the surface of a sphere of radius R, say, centered at the origin. Since $d\boldsymbol{S} = \hat{\boldsymbol{r}} dS$, we have

$$\oint_{S'} \boldsymbol{u} \cdot d\boldsymbol{S} = \frac{\Lambda}{4\pi R^2}\oint_{S'} dS = \Lambda,$$

since the surface area of the sphere is $4\pi R^2$. Therefore,

$$\int_V \boldsymbol{\nabla} \cdot \boldsymbol{u}\, dV = \begin{cases} 0, & (0,0,0) \notin V; \\ \Lambda, & (0,0,0) \in V. \end{cases}$$

For those of you familiar with the one-dimensional Dirac delta function, say from my course *Differential Equations for Engineers*, what we have here is

$$\boldsymbol{\nabla} \cdot \boldsymbol{u} = \Lambda\delta(\boldsymbol{r}),$$

where $\delta(\boldsymbol{r})$ is the three-dimensional Dirac delta function satisfying

$$\delta(\boldsymbol{r}) = 0, \qquad \text{when } \boldsymbol{r} \neq 0,$$

and

$$\int_V \delta(\boldsymbol{r})\, dV = 1, \qquad \text{provided the origin is in } V.$$

In Cartesian or in spherical coordinates, the three-dimensional Dirac delta function centered at the origin may be written as

$$\delta(\boldsymbol{r}) = \delta(x)\delta(y)\delta(z) = \frac{1}{4\pi r^2}\delta(r),$$

where $\delta(x)$, $\delta(y)$, $\delta(z)$ and $\delta(r)$ are one-dimensional Dirac delta functions.

Solutions to the Problems for Lecture 49

1. The continuity equation as derived in the lecture is given by

$$\frac{\partial \rho}{\partial t} + \boldsymbol{\nabla} \cdot (\rho \boldsymbol{u}) = 0.$$

Using the vector identity $\boldsymbol{\nabla} \cdot (\rho \boldsymbol{u}) = \boldsymbol{u} \cdot \boldsymbol{\nabla}\rho + \rho \boldsymbol{\nabla} \cdot \boldsymbol{u}$, the continuity equation becomes

$$\frac{\partial \rho}{\partial t} + \boldsymbol{u} \cdot \boldsymbol{\nabla}\rho + \rho \boldsymbol{\nabla} \cdot \boldsymbol{u} = 0.$$

2. We begin with

$$\frac{d}{dt}\int_V \rho(\boldsymbol{r}, t)\, dV = -\oint_S \boldsymbol{J} \cdot d\boldsymbol{S}.$$

The divergence theorem applied to the right-hand side results in

$$\oint_S \boldsymbol{J} \cdot d\boldsymbol{S} = \int_V \boldsymbol{\nabla} \cdot \boldsymbol{J}\, dV;$$

and combining both sides of the equation and bringing the time derivative inside the integral results in

$$\int_V \left(\frac{\partial \rho}{\partial t} + \boldsymbol{\nabla} \cdot \boldsymbol{J}\right) dV = 0.$$

Since the integral is zero for any volume V, we obtain the electrodynamics continuity equation given by

$$\frac{\partial \rho}{\partial t} + \boldsymbol{\nabla} \cdot \boldsymbol{J} = 0.$$

Solutions to the Practice quiz: Divergence theorem

1. a. With $\boldsymbol{u} = yz\boldsymbol{i} + xz\boldsymbol{j} + xy\boldsymbol{k}$, we have $\boldsymbol{\nabla} \cdot \boldsymbol{u} = 0$. Therefore,

$$\oint_S \boldsymbol{u} \cdot d\boldsymbol{S} = \int_V (\boldsymbol{\nabla} \cdot \boldsymbol{u})\, dV = 0.$$

2. d.

$$\oint_S \boldsymbol{r} \cdot d\boldsymbol{S} = \int_V (\boldsymbol{\nabla} \cdot \boldsymbol{r})\, dV = 3\int_V dV = 3\pi R^2 L.$$

3. d. Computing the divergences, we have

$$\begin{aligned}
&\boldsymbol{\nabla} \cdot \left[xy\boldsymbol{i} - \frac{1}{2}y^2\boldsymbol{j}\right] = y - y = 0,\\
&\boldsymbol{\nabla} \cdot [(1+x)\boldsymbol{i} + (1-y)\boldsymbol{j}] = 1 - 1 = 0,\\
&\boldsymbol{\nabla} \cdot \left[(x^2 - xy)\boldsymbol{i} + \left(\frac{1}{2}y^2 - 2xy\right)\boldsymbol{j}\right] = (2x - y) + (y - 2x) = 0,\\
&\boldsymbol{\nabla} \cdot \left[(x+y)^2\boldsymbol{i} + (x-y)^2\boldsymbol{j}\right] = 2(x+y) - 2(x-y) = 4y.
\end{aligned}$$

Solutions to the Problems for Lecture 50

1. With $\boldsymbol{u} = -y\boldsymbol{i} + x\boldsymbol{j}$, we use $\partial u_2/\partial x - \partial u_1/\partial y = 2$. For a square of side L, we have for the left-hand side of Green's theorem

$$\int_A \left(\frac{\partial u_2}{\partial x} - \frac{\partial u_1}{\partial y}\right) dA = 2\int_A dA = 2L^2.$$

When the square lies in the first quadrant with vertex at the origin, we have for the right-hand side of Green's theorem,

$$\oint_C (u_1\, dx + u_2\, dy) = \int_0^L 0\, dx + \int_L^0 (-L)\, dx + \int_L^0 0\, dy + \int_0^L L\, dy = 2L^2.$$

2. With $\boldsymbol{u} = -y\boldsymbol{i} + x\boldsymbol{j}$, we use $\partial u_2/\partial x - \partial u_1/\partial y = 2$. For a circle of radius R, we have for the left-hand side of Green's theorem,

$$\int_A \left(\frac{\partial u_2}{\partial x} - \frac{\partial u_1}{\partial y}\right) dA = 2\int_A dA = 2\pi R^2.$$

For a circle of radius R centered at the origin, we change variables to $x = R\cos\theta$ and $y = R\sin\theta$. Then $dx = -R\sin\theta\, d\theta$ and $dy = R\cos\theta\, d\theta$, and we have for the right-hand side of Green's theorem,

$$\oint_C (u_1\, dx + u_2\, dy) = \oint_C (-y\, dx + x\, dy) = \int_0^{2\pi} (R^2 \sin^2\theta + R^2\cos^2\theta) d\theta = 2\pi R^2.$$

Solutions to the Problems for Lecture 51

1. Let $\boldsymbol{u} = u_1(x,y,z)\,\boldsymbol{i} + u_2(x,y,z)\,\boldsymbol{j} + u_3(x,y,z)\,\boldsymbol{k}$. Then

$$\boldsymbol{\nabla} \times \boldsymbol{u} = \left(\frac{\partial u_3}{\partial y} - \frac{\partial u_2}{\partial z}\right)\boldsymbol{i} + \left(\frac{\partial u_1}{\partial z} - \frac{\partial u_3}{\partial x}\right)\boldsymbol{j} + \left(\frac{\partial u_2}{\partial x} - \frac{\partial u_1}{\partial y}\right)\boldsymbol{k}.$$

a) For an area lying in the y-z plane bounded by a curve C, the normal vector to the area is $\boldsymbol{i}$. Therefore, Green's theorem is given by

$$\int_A \left(\frac{\partial u_3}{\partial y} - \frac{\partial u_2}{\partial z}\right) dA = \oint_C (u_2 dy + u_3 dz);$$

b) For an area lying in the z-x plane bounded by a curve C, the normal vector to the area is $\boldsymbol{j}$. Therefore, Green's theorem is given by

$$\int_A \left(\frac{\partial u_1}{\partial z} - \frac{\partial u_3}{\partial x}\right) dA = \oint_C (u_3 dz + u_1 dx);$$

The correct orientation of the curves are determined by the right-hand rule, using a right-handed coordinate system.

2. We have $\boldsymbol{u} = -y\boldsymbol{i} + x\boldsymbol{j}$. The right-hand side of Stokes' theorem was computed in an earlier problem on Green's theorem and we repeat the solution here. For a circle of radius R lying in the x-y plane with center at the origin, we change variables to $x = R\cos\phi$ and $y = R\sin\phi$. Then $dx = -R\sin\phi\, d\phi$ and $dy = R\cos\phi\, d\phi$, and we have for the right-hand side of Stokes' theorem,

$$\begin{aligned}\oint_C \boldsymbol{u}\cdot d\boldsymbol{r} = \oint_C (u_1\, dx + u_2\, dy) &= \oint_C (-y\, dx + x\, dy) \\ &= \int_0^{2\pi} (R^2\sin^2\phi + R^2\cos^2\phi) d\phi = 2\pi R^2.\end{aligned}$$

The left-hand side of Stokes' theorem uses

$$\nabla \times \boldsymbol{u} = \begin{vmatrix} \boldsymbol{i} & \boldsymbol{j} & \boldsymbol{k} \\ \partial/\partial x & \partial/\partial y & \partial/\partial z \\ -y & x & 0 \end{vmatrix} = 2\boldsymbol{k};$$

so that with $d\boldsymbol{S} = \hat{\boldsymbol{r}} R^2 \sin\theta \, d\theta \, d\phi$, we have

$$\int_S (\nabla \times \boldsymbol{u}) \cdot d\boldsymbol{S} = 2R^2 \int_0^{2\pi} \int_0^{\pi/2} \boldsymbol{k} \cdot \hat{\boldsymbol{r}} \sin\theta \, d\theta \, d\phi.$$

With

$$\boldsymbol{k} = \cos\theta \, \hat{\boldsymbol{r}} - \sin\theta \, \hat{\boldsymbol{\theta}},$$

we have

$$\boldsymbol{k} \cdot \hat{\boldsymbol{r}} = \cos\theta;$$

and

$$\begin{aligned} \int_S (\nabla \times \boldsymbol{u}) \cdot d\boldsymbol{S} &= 2R^2 \int_0^{2\pi} d\phi \int_0^{\pi/2} \cos\theta \sin\theta \, d\theta \\ &= 2\pi R^2 \sin^2\theta \Big|_0^{\pi/2} = 2\pi R^2. \end{aligned}$$

3. We consider the two-dimensional velocity field given by

$$\boldsymbol{u} = \frac{\Gamma}{2\pi} \left(\frac{-y\boldsymbol{i} + x\boldsymbol{j}}{x^2 + y^2} \right).$$

a) Cylindrical coordinates are defined by

$$x = \rho\cos\phi, \quad y = \rho\sin\phi, \quad \boldsymbol{i} = \cos\phi\hat{\boldsymbol{\rho}} - \sin\phi\hat{\boldsymbol{\phi}}, \quad \boldsymbol{j} = \sin\phi\hat{\boldsymbol{\rho}} + \cos\phi\hat{\boldsymbol{\phi}}.$$

Substituting into $\boldsymbol{u}$, we obtain

$$\boldsymbol{u} = \frac{\Gamma\hat{\boldsymbol{\phi}}}{2\pi\rho}.$$

b) Let $\boldsymbol{\omega} = \nabla \times \boldsymbol{u}$. For $\rho \neq 0$, our velocity field satisfies $u_\rho = 0$, $u_\phi = \Gamma/(2\pi\rho)$, $u_z = 0$, and we find using cylindrical coordinates,

$$\boldsymbol{\omega} = \nabla \times \boldsymbol{u} = \hat{\boldsymbol{k}} \left(\frac{1}{\rho}\frac{\partial}{\partial\rho}(\rho u_\phi) - \frac{1}{\rho}\frac{\partial u_\rho}{\partial\phi} \right) = \hat{\boldsymbol{k}} \left(\frac{1}{\rho}\frac{\partial}{\partial\rho}\left(\frac{\Gamma}{2\pi}\right) \right) = 0.$$

c) We now consider a surface integral of the vorticity field over an area in the x-y plane containing the origin, i.e.,

$$\int_S \boldsymbol{\omega} \cdot \hat{\boldsymbol{k}} dS.$$

Because $\boldsymbol{\omega} = 0$ everywhere except at $\rho = 0$, we can reduce this integral to be that over a circle of radius R centered at the origin. Then applying Stokes' theorem,

$$\int_S \boldsymbol{\omega} \cdot \hat{\boldsymbol{k}} dS = \int_S (\boldsymbol{\nabla} \times \boldsymbol{u}) \cdot \hat{\boldsymbol{k}} dS = \oint_C \boldsymbol{u} \cdot d\boldsymbol{r} = \int_0^{2\pi} \left(\frac{\Gamma \hat{\boldsymbol{\phi}}}{2\pi R} \right) \cdot \hat{\boldsymbol{\phi}} R d\phi = \Gamma.$$

Now $\boldsymbol{\omega}$ equals zero everywhere except at the origin, and the two-dimensional integral of its z-component over any area in the x-y plane containing the origin is equal to Γ. We can therefore identify the z-component of $\boldsymbol{\omega}$ with Γ times the two-dimensional Dirac delta function, i.e.,

$$\boldsymbol{\omega} = \Gamma \delta(x) \delta(y) \hat{\boldsymbol{k}},$$

where $\delta(x)$ and $\delta(y)$ are one-dimensional Dirac delta functions. This vorticity field is called a two-dimensional point vortex. The two-dimensional Dirac delta function, here written in Cartesian coordinates, can also be written in cylindrical (or polar) coordinates as

$$\delta(x)\delta(y) = \frac{1}{2\pi\rho} \delta(\rho).$$

Solutions to the Practice quiz: Stokes' theorem

1. b. With $\boldsymbol{u} = -y\boldsymbol{i} + x\boldsymbol{j}$, we have $\boldsymbol{\nabla} \times \boldsymbol{u} = 2\boldsymbol{k}$ and we use Stokes' theorem to write

$$\oint_C \boldsymbol{u} \cdot d\boldsymbol{r} = \int_S (\boldsymbol{\nabla} \times \boldsymbol{u}) \cdot d\boldsymbol{S} = 2 \int dA = \frac{1}{2} \pi R^2,$$

where we have used $d\boldsymbol{S} = \boldsymbol{k}\, dA$ and the area of the quarter circle is $\frac{1}{4}\pi R^2$.

2. c. With $\boldsymbol{u} = \dfrac{-y}{x^2+y^2}\boldsymbol{i} + \dfrac{x}{x^2+y^2}\boldsymbol{j}$, one can show by differentiating that $\boldsymbol{\nabla} \times \boldsymbol{u} = 0$ provided $(x, y) \neq (0, 0)$. However, the integration region contains the origin so the integral is best done by applying Stokes' theorem. We use cylindrical coordinates to write

$$\boldsymbol{u} = \frac{-y}{x^2+y^2}\boldsymbol{i} + \frac{x}{x^2+y^2}\boldsymbol{j} = \frac{\hat{\boldsymbol{\phi}}}{\rho}.$$

Then,

$$\int_S (\boldsymbol{\nabla} \times \boldsymbol{u}) \cdot d\boldsymbol{S} = \oint_C \boldsymbol{u} \cdot d\boldsymbol{r} = \int_0^{2\pi} \left(\frac{\hat{\boldsymbol{\phi}}}{\rho}\right) \cdot (\hat{\boldsymbol{\phi}}\rho d\phi) = \int_0^{2\pi} d\phi = 2\pi.$$

3. c. With $\boldsymbol{u} = -x^2y\boldsymbol{i} + xy^2\boldsymbol{j}$, we have $\boldsymbol{\nabla} \times \boldsymbol{u} = (x^2 + y^2)\boldsymbol{k}$. Therefore, with $d\boldsymbol{S} = \boldsymbol{k}dxdy$, we have

$$\begin{aligned}\oint_C \boldsymbol{u} \cdot d\boldsymbol{r} = \int_S (\boldsymbol{\nabla} \times \boldsymbol{u}) \cdot d\boldsymbol{S} &= \int_0^1 \int_0^1 (x^2 + y^2)\, dx\, dy \\ &= \int_0^1 x^2\, dx \int_0^1 dy + \int_0^1 dx \int_0^1 y^2\, dy = \frac{2}{3}.\end{aligned}$$

Solutions to the Problems for Lecture 52

1.

a) The Navier-Stokes equation and the continuity equation are given by

$$\frac{\partial \boldsymbol{u}}{\partial t} + (\boldsymbol{u} \cdot \boldsymbol{\nabla})\boldsymbol{u} = -\frac{1}{\rho}\boldsymbol{\nabla} p + \nu \nabla^2 \boldsymbol{u}, \qquad \boldsymbol{\nabla} \cdot \boldsymbol{u} = 0.$$

Taking the divergence of both sides of the Navier-Stokes equation and using the continuity equation results in

$$\boldsymbol{\nabla} \cdot ((\boldsymbol{u} \cdot \boldsymbol{\nabla})\boldsymbol{u}) = -\frac{1}{\rho}\nabla^2 p.$$

Now,

$$\boldsymbol{\nabla} \cdot ((\boldsymbol{u} \cdot \boldsymbol{\nabla})\boldsymbol{u}) = \frac{\partial}{\partial x_i}\left(u_j \frac{\partial u_i}{\partial x_j}\right) = \frac{\partial u_i}{\partial x_j}\frac{\partial u_j}{\partial x_i}.$$

Therefore,

$$\nabla^2 p = -\rho \frac{\partial u_i}{\partial x_j}\frac{\partial u_j}{\partial x_i}.$$

b) Taking the curl of both sides of the Navier-Stokes equation, and using $\boldsymbol{\omega} = \boldsymbol{\nabla} \times \boldsymbol{u}$ and $\boldsymbol{\nabla} \times \boldsymbol{\nabla} p = 0$, we obtain

$$\frac{\partial \boldsymbol{\omega}}{\partial t} + \boldsymbol{\nabla} \times (\boldsymbol{u} \cdot \boldsymbol{\nabla})\boldsymbol{u} = \nu \nabla^2 \boldsymbol{\omega}.$$

To simplify the second term, we first prove the identity

$$\boldsymbol{u} \times (\boldsymbol{\nabla} \times \boldsymbol{u}) = \frac{1}{2}\boldsymbol{\nabla}(\boldsymbol{u} \cdot \boldsymbol{u}) - (\boldsymbol{u} \cdot \boldsymbol{\nabla})\boldsymbol{u}.$$

We prove by considering the ith component of the left-hand side:

$$\begin{aligned}
[\boldsymbol{u} \times (\boldsymbol{\nabla} \times \boldsymbol{u})]_i &= \epsilon_{ijk} u_j \epsilon_{klm} \frac{\partial u_m}{\partial x_l} = \epsilon_{kij}\epsilon_{klm} u_j \frac{\partial u_m}{\partial x_l} \\
&= (\delta_{il}\delta_{jm} - \delta_{im}\delta_{jl}) u_j \frac{\partial u_m}{\partial x_l} = u_j \frac{\partial u_j}{\partial x_i} - u_j \frac{\partial u_i}{\partial x_j} \\
&= \frac{1}{2}\frac{\partial}{\partial x_i}(u_j u_j) - u_j \frac{\partial u_i}{\partial x_j} \\
&= \left[\frac{1}{2}\boldsymbol{\nabla}(\boldsymbol{u} \cdot \boldsymbol{u})\right]_i - [(\boldsymbol{u} \cdot \boldsymbol{\nabla})\boldsymbol{u}]_i.
\end{aligned}$$

Therefore, using the facts that the curl of a gradient and the divergence of a curl are equal to zero, and $\boldsymbol{\omega} = \boldsymbol{\nabla} \times \boldsymbol{u}$ and $\boldsymbol{\nabla} \cdot \boldsymbol{u} = 0$, we have

$$\begin{aligned}
\boldsymbol{\nabla} \times (\boldsymbol{u} \cdot \boldsymbol{\nabla})\boldsymbol{u} &= \boldsymbol{\nabla} \times \left(\frac{1}{2}\boldsymbol{\nabla}(\boldsymbol{u} \cdot \boldsymbol{u}) - \boldsymbol{u} \times (\boldsymbol{\nabla} \times \boldsymbol{u})\right) \\
&= -\boldsymbol{\nabla} \times (\boldsymbol{u} \times \boldsymbol{\omega}) \\
&= -\left[\boldsymbol{u}(\boldsymbol{\nabla} \cdot \boldsymbol{\omega}) - \boldsymbol{\omega}(\boldsymbol{\nabla} \cdot \boldsymbol{u}) + (\boldsymbol{\omega} \cdot \boldsymbol{\nabla})\boldsymbol{u} - (\boldsymbol{u} \cdot \boldsymbol{\nabla})\boldsymbol{\omega}\right] \\
&= -(\boldsymbol{\omega} \cdot \boldsymbol{\nabla})\boldsymbol{u} + (\boldsymbol{u} \cdot \boldsymbol{\nabla})\boldsymbol{\omega}.
\end{aligned}$$

Putting it all together gives us the vorticity equation, given by

$$\frac{\partial \boldsymbol{\omega}}{\partial t} + (\boldsymbol{u} \cdot \boldsymbol{\nabla})\boldsymbol{\omega} = (\boldsymbol{\omega} \cdot \boldsymbol{\nabla})\boldsymbol{u} + \nu \nabla^2 \boldsymbol{\omega}.$$

Solutions to the Problems for Lecture 53

1. The electric field from a point charge at the origin should be spherically symmetric. We therefore write using spherical coordinates, $\boldsymbol{E}(\boldsymbol{r}) = E(r)\hat{\boldsymbol{r}}$. Applying Gauss's law to a spherical shell of radius r, we have

$$\oint_S \boldsymbol{E} \cdot d\boldsymbol{S} = E(r) \oint_S dS = 4\pi r^2 E(r) = \frac{q}{\varepsilon_0}.$$

Therefore, the electric field is given by

$$\boldsymbol{E}(\boldsymbol{r}) = \frac{q}{4\pi\varepsilon_0 r^2}\hat{\boldsymbol{r}}.$$

2. The magnetic field from a current carrying infinite wire should have cylindrical symmetry. We therefore write using cylindrical coordinates, $\boldsymbol{B}(\boldsymbol{r}) = B(\rho)\hat{\boldsymbol{\phi}}$. Applying Ampère's law to a circle of radius ρ in the x-y plane in the counterclockwise direction, we obtain

$$\oint_C \boldsymbol{B} \cdot d\boldsymbol{r} = B(\rho) \oint_C dr = 2\pi\rho B(\rho) = \mu_0 I,$$

where I is the current in the wire. Therefore,

$$\boldsymbol{B}(\boldsymbol{r}) = \frac{\mu_0 I}{2\pi\rho}\hat{\boldsymbol{\phi}}.$$

Made in the USA
Coppell, TX
24 August 2024

36414373R00125